Do Your Own
HOME WIRING

DO YOUR OWN
HOME WIRING

GEOFFREY BURDETT

London
W. FOULSHAM & CO LTD
New York . Toronto . Capetown . Sydney

W. FOULSHAM & CO. LTD
Yeovil Road, Slough, Berks, England

Revised and updated 1985

ISBN 0–572–01327–2

Printed in Great Britain at
The Bath Press, Avon

CONTENTS

ACKNOWLEDGMENTS

The Author and Publishers wish to thank the following companies who have kindly given permission for photographs to be reproduced.

Belling & Co. Ltd.
Black and Decker Ltd.
British Home Stores
E. Chidlow & Co. Ltd.
Coventry Controls Ltd.
J. A. Crabtree & Co. Ltd.
Creda Electric Ltd.
Findlay, Durham & Brodie
Fotherby Willis Electronics Ltd.
Gardom & Lock Ltd.
Haffenden-Richborough Ltd.
Heatrae Ltd.
Humex Ltd.
I. M. I. Santon Ltd.
Midland Electric Manufacturing Co. Ltd.
MK Electric Ltd.
Nettle Accessories Ltd.
Ottermill Switchgear Ltd.
P. & R. Electrical (London) Ltd.
Philips Electrical Ltd.
Qualcast Ltd.
Rock Electrical Accessories Ltd.
George H. Scholes & Co. Ltd.
Stanley-Bridges Ltd.
Thorn Lighting Ltd.

INTRODUCTION

The electricity supply system in the UK is standardized at 240/415V, and is a.c. (alternating current) at a standard frequency of 50Hz, which is the metric equivalent of the former 50 cycles per second.

A four-wire distribution cable runs along the street (Figure 1). Three cores of this cable are phase wires; the fourth is the neutral, the supply system being three-phase and neutral. Voltage between any two phases is 415V and between any one phase and the neutral, 240V.

The neutral conductor is connected to earth within the system. This means that the voltage between any one phase and the earth is the same as between any one phase and the neutral – which is 240V. The 'voltage' between the neutral and earth is therefore zero, the neutral being referred to as the earthed neutral.

Most dwellings are supplied from one phase of the system only, and are said to receive a single-phase supply. However, where an above-normal load in the form of electrical appliances is installed and the electricity demand is likely to be substantial, a dwelling will usually be supplied from more than one phase of the system. The electrical load in the house is then evenly divided over the different phases.

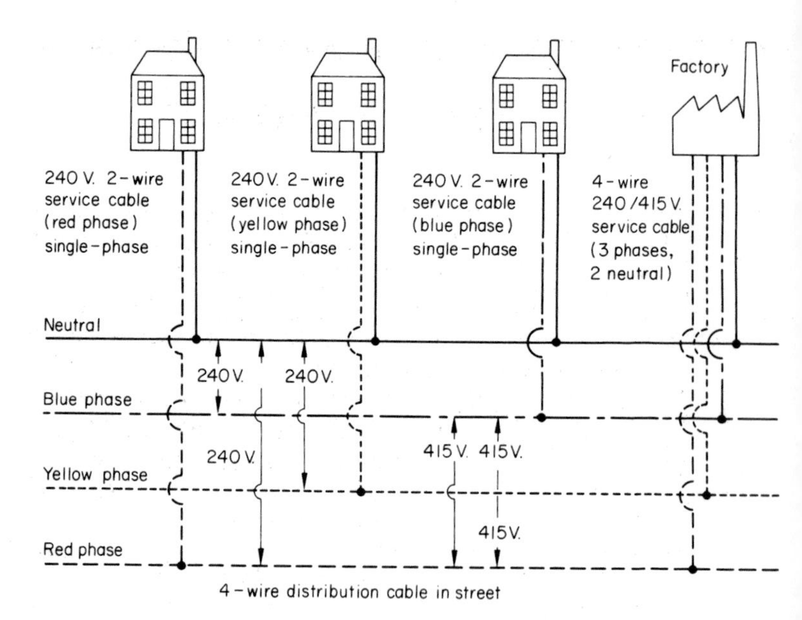

Fig. 1. Electricity distribution system supplying houses on single-phase and factories and offices on three-phase.

1

ELECTRICITY INTO THE HOUSE

Electricity comes into a house via an underground cable in cities and towns and via overhead cables in villages and rural areas. These service cables have two cores: one is the phase or live wire; the other is the earthed neutral (Fig. 2). Where more than one phase is supplied to a house the service cable has an additional core for each additional phase.

The service cable terminates at a connection box, which contains a service fuse (or fuses, if there is more than one phase) and a solid link for the neutral. Fuses may be inserted only in a live pole, never in the earthed neutral. Adjacent to the service

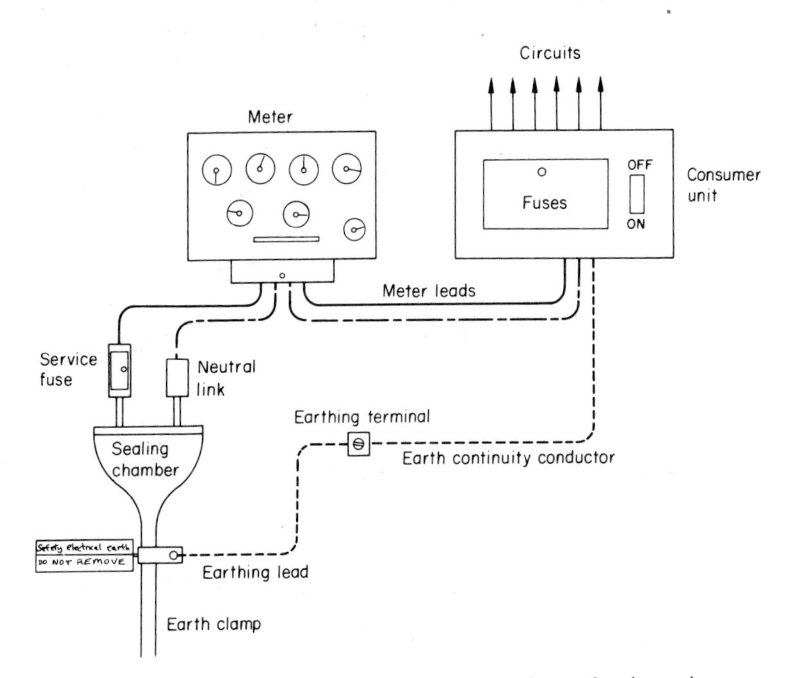

Fig. 2. Typical arrangement at the supply intake position in a house showing mains cable, service fuse and meter.

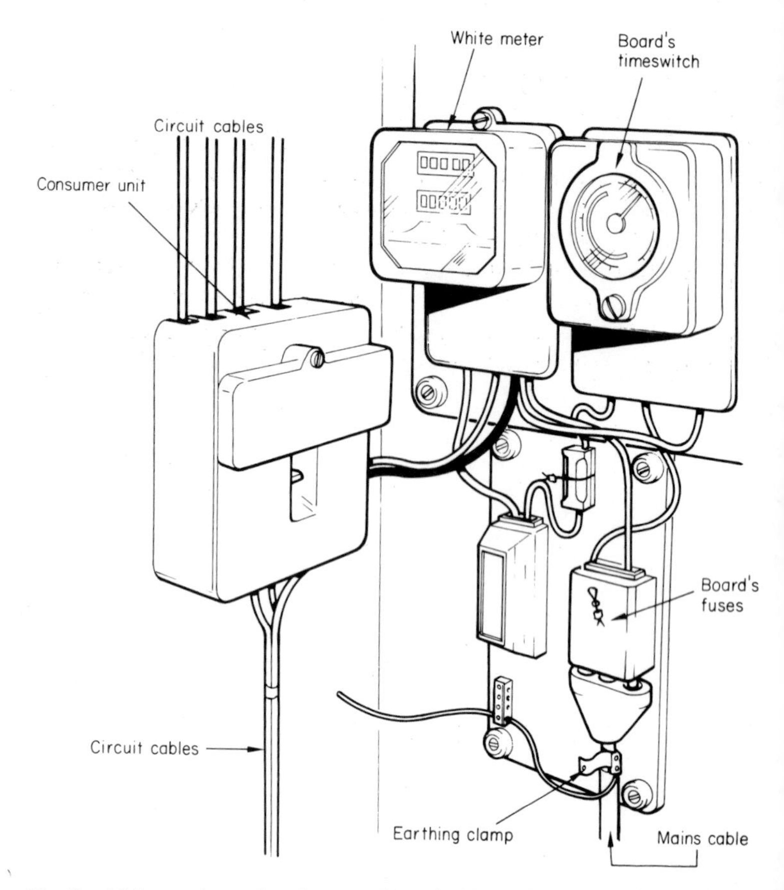

Fig. 3. Mains service and equipment of a typical home installation which includes "White Meter", time switch, Board's cable, fuses, and terminal box and consumer unit.

connection unit is the electricity meter. Both are mounted on one board and are linked together by two substantial sheathed cables. Both the meter and the service unit are sealed by the electricity board and remain the property of the board (Fig. 3).

Where the householder has chosen the Economy 7 tariff (called White Meter in Scotland) a white meter with two rows of figures is installed. The top row shows the units used at the cheap rate charged during 7 hours of the night and the bottom row shows the units used at the normal day rate. A time switch changes the rate at the appropriate times.

COLOUR CODE

In order to distinguish the wires from each other, the insulation of the live wire is red, that of the neutral wire is black and the earth wire is either bare or covered with a green/yellow striped sleeve in accessories where it could come in contact with the live or neutral terminals.

Flexes used for lights and portable equipment have their wire insulation covered brown for live, blue for neutral and yellow/green striped for earth.

2

THE CONSUMER'S INSTALLATION

All wiring and equipment on the house side of the meter is the responsibility of the householder and is termed the consumer's installation. This is connected to the meter by two substantial cables of the same type and usually of the same size as those connecting the meter to the service unit.

The connection at the consumer's side is made at an obligatory double-pole main switch, fixed within reach of the board's service fuse and meter. In the modern installation this switch is embodied in a composite consumer unit which also contains the necessary circuit fuses. Old installations which have a number of main switches and fuse units, now need to be rewired and re-equipped.

THE CONSUMER UNIT

The consumer unit, as stated, is a combined double-pole main switch and fuse distribution board. It is made in a number of sizes to suit the installation. The smallest size is a two-way unit containing two circuit fuses. The largest installed in the home is a ten-way unit with ten circuit fuses. One consumer unit supplies all circuits of the home installation with the exception of electric storage heating, which has to be supplied from another consumer unit under a time switch control.

The size of consumer unit is determined by the number of circuits comprising the home installation. Each circuit requires a separate fuse way. If for example there are six circuits, a six-way consumer unit is needed. This is the average size fitted in the conventional three-bedroom house (Fig. 4). It is however wise to include at least one extra fuse way as a spare for possible future extensions to the installation.

Fuse units, and the fuses that fit them (Fig. 5), have various current ratings according to the circuits they supply (see table 1). There are four standard ratings, 5A, 15A, 20A and 30A, but some consumer units have facilities for a 45A fuse way, to supply circuits for the larger-size family electric cooker.

Fuse units of the four standard current ratings are interchangeable within the consumer unit. This means that a consumer unit can contain any mixture of ratings to suit installation requirements. The fuses may be placed in any

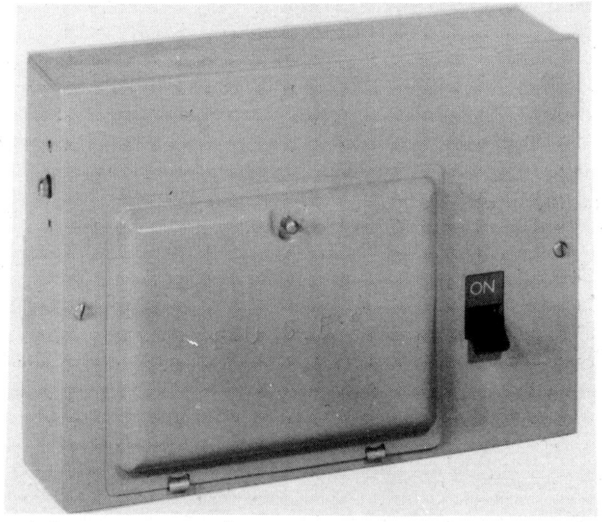

Fig. 4. A six-way consumer unit.

Fig. 5. A six-way consumer unit showing fuses.

order of ratings, but preferably the fuse of the largest current rating is positioned next to the main switch and the remainder in descending order so that those of the smallest current rating (5A) are farthest away from the main switch. Fuse ways are in fact numbered from the main switch, although in some makes the main switch is on the right of the fuse ways, while in others it is on the left.

Fuse units and their fuses are colour-coded (table 1), and fuse holders (the removable section containing the fuse wire or fuse cartridge) of the different current ratings have different physical dimensions. The colour indicates the fuse rating and the different physical dimensions prevent, say, a 30A fuse being fitted into a fuse unit of lower rating. Fuses are of two types: rewirable fuses containing a fuse wire element (Fig. 6), and cartridge fuses, made in the form of a cartridge which totally encloses the fuse element, preventing it being rewired. Most consumer units in use have rewirable fuses but the cartridge fuse is the more reliable though more expensive. A disadvantage of cartridge fuses is that spare cartridges of each current rating must be available in the same way that a card of fuse wire should be available for mending rewirable fuses.

Table 1 Circuit fuses

Rating (amps)	Colour code	Circuit protected
5	White	Lighting
15	Blue	Immersion heaters and other 3kW circuits
20	Yellow	Multi-socket radial circuit
30	Red	Ring circuits; small cooker; 5–7kW water heater; radial circuit; supply to detached building
45	Green	Large cooker

An alternative to the circuit fuse is the miniature circuit breaker (m.c.b.) and this is fitted in some models of consumer unit instead of fuses. These have current ratings to suit the circuits (see Figs. 7 & 8). When a fault occurs in a circuit the m.c.b. switches off automatically. To restore the current to the circuit the m.c.b. is switched on again, but if the fault persists it becomes impossible to 'close' the circuit breaker until the fault is rectified.

M.c.b.s are superior to fuses. They are much more expensive to buy, but once fitted there will be no fuses to mend in the future.

Fig. 6. Rewirable fuse unit, cartridge fuse unit and various fuses.

INSTALLING A CONSUMER UNIT

Select a consumer unit of the required size (number of fuse ways) and type, fitted with rewirable or cartridge fuses or with m.c.b.s of the required current ratings for the various circuits.

Prepare the wall next to the meter for fixing the unit. If the consumer unit is replacing a miscellany of switch fuse units and other switch and fuse gear, these must first be disconnected from the mains and removed *after the electricity board has withdrawn the service fuse to cut off the current.*

Fig. 7. Consumer unit with m.c.b.s.

Fig. 8. Miniature circuit breaker (m.c.b.)

Remove the consumer unit cover and take out all the fuse units or m.c.b.s.

If the consumer unit has a plastic casing it will probably have an open back for mounting over a conduit junction box of a conduit installation. For a PVC-sheathed installation an insulated and non-combustible backing sheet is supplied with the unit for insertion between the consumer unit and the wall and is fixed with it. Metal-cased units will be totally enclosed and the circuit cables will pass through cable entry holes fitted with PVC or rubber grommets.

Having dealt with these preliminaries, fix the unit to the wall using wood screws in plugged holes in the wall.

Thread in the circuit cables and cut each to its approximate length, marking the circuit of each. Strip off the sheath of each cable in turn, leaving about 50mm of sheath within the unit. Check the conductors for the final lengths and strip off about 12mm of insulation for insertion in the terminals.

As you strip each sheathed cable, connect the conductors to the respective fuse ways. The red insulated conductor is secured in the fuse terminal and the black in the corresponding terminal hole in the neutral terminal block. Enclose the bare earth continuity conductor in green yellow PVC sleeving, leaving 12mm for the connection in the earth terminal block. For each ring circuit there are two cables, so two red insulated conductors are inserted in one fuse way terminal, two black insulated conductors in the corresponding neutral terminal hole, and two earth conductors in the corresponding earth terminal hole.

Having connected all the circuit cables and left the spare fuse way blank, now connect the meter leads to the main switch terminals. Strip the ends by removing about 38mm of sheathing and 25mm of insulation. Secure the red insulated conductor to the 'L' terminal and the black to the 'N' terminal. Fit the insulated shields (if supplied) over the mains terminals and coil the cables out of the way for the electricity board to connect to the meter. These cables should be not less than 1m long, depending on the distance of the meter from the consumer unit. The earthing lead is connected to the remaining hole of the earth terminal block and is coiled up out of the way.

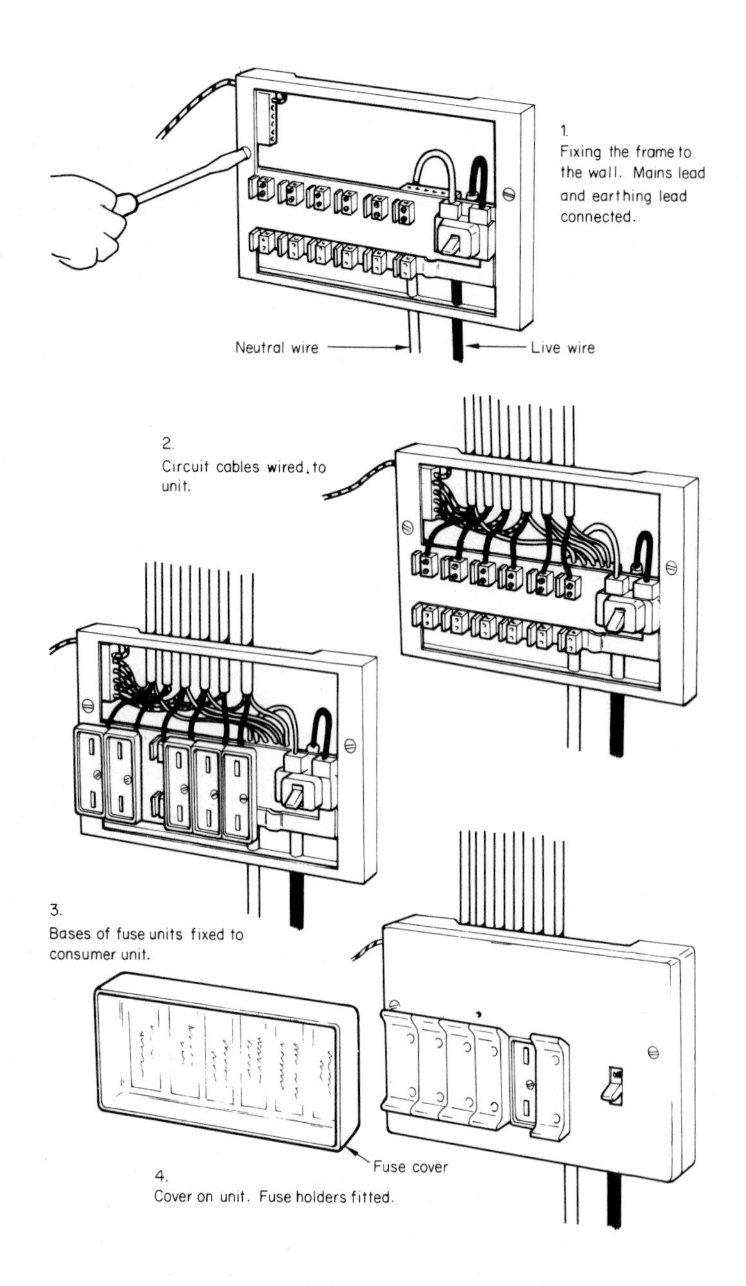

Fig. 9. Installing a consumer unit.

Now fix the fuse units and the insulated terminal cover plate. Insert the fuses and fit the outer cover and the detachable fuse cover, having first inserted details of the circuits.

Where the consumer unit has m.c.b.s instead of fuses these are fitted before the circuit conductors are connected, because the conductors are connected to terminals on the m.c.b.s. M.c.b.s are not enclosed in a cover but are exposed to give access to the switches. With some the circuit details are entered on labels on the m.c.b.s.

The end of the PVC-insulated earthing wire is connected to an earthing terminal situated near the meter. If an earth leakage circuit breaker is installed, however, there is no earthing terminal. Instead, the earthing lead either goes to an earth electrode rod outside the house or is connected to the 'F' terminal on the circuit breaker depending on whether the circuit breaker is current-operated or voltage-operated (see Fig. 10).

The consumer unit is now ready for connection by the electricity board following the completion of an application form.

EARTH LEAKAGE CIRCUIT BREAKERS

An earth leakage circuit breaker (e.l.c.b.) is a form of main switch containing a trip coil which, when energized by a current leakage between a live conductor and earthed metal, trips the main switch and cuts off the current from the mains (Fig. 11).

A fuse normally does this, but it requires a very large current to flow through the earthing system to the electricity substation, which may be a mile or so away. For example, 60A of current has to flow through the fault and through the earthing system to blow a 30A rewirable fuse. An e.l.c.b., however, will operate when only 1A or even less flows through the earthing system.

Normally the earthing system of an installation is connected (earthed) to the metal sheath of the electricity board's underground cable, which effectively carries large currents back to the sub-station and in so doing blows the fuse; but in some instances the board uses other methods. Also, until recently an installation could be connected to the mains water pipe, which has sufficient metal buried in the earth for the earth itself to

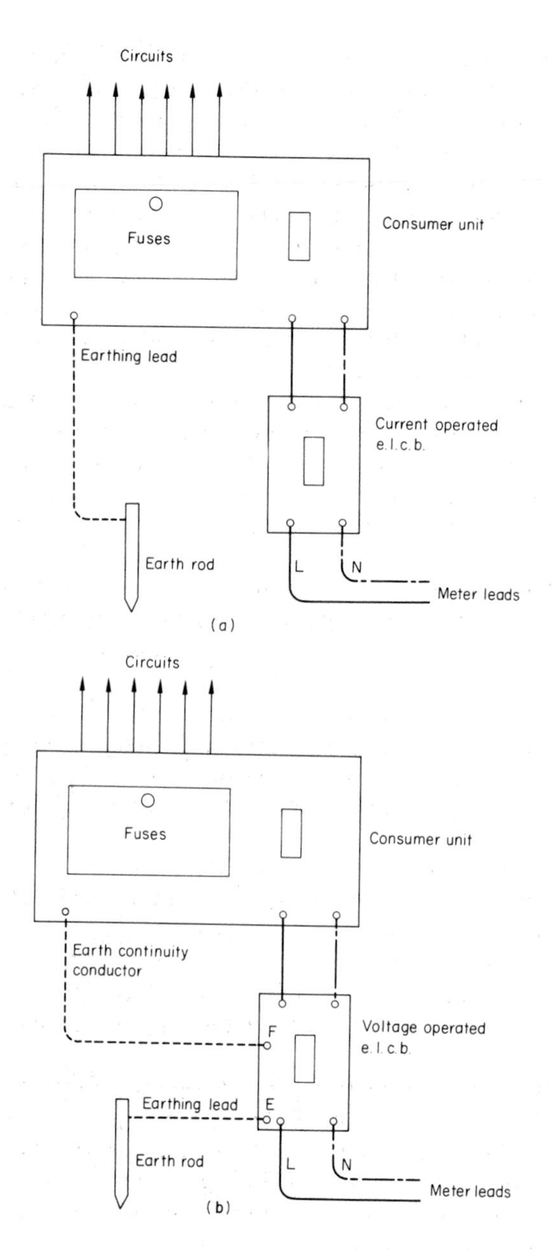

Fig. 10. Wiring connections for the two types of earth leakage circuit breaker: a) current-operated; b) voltage-operated.

carry the current. However, with the introduction of plastic pipes this is no longer regarded as an effective earth.

Where the electricity board is unable to provide earthing facilities, and mains water pipes are partly or entirely plastic, the householder must provide his own earthing facilities, for earthing is the responsibility of the consumer, and not of the board.

The obvious means of doing this is an earth electrode, a purpose-made copper-coated steel rod of 1.2m or more in length driven into the garden soil. This by itself is rarely satisfactory, however, for it will not pass large currents into the soil, especially during dry weather when the resistance of the soil is very high. As a result, e.l.c.b.s are installed to operate in conjunction with earth rod electrodes.

Of the two types, current-operated and voltage-operated,

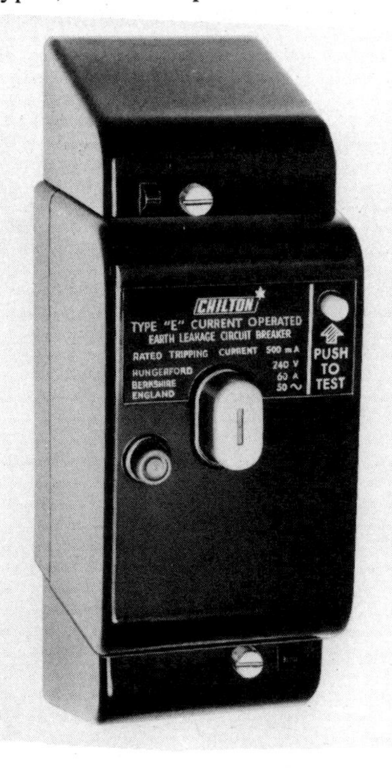

Fig. 11. Earth leakage circuit breaker (e.l.c.b.)

the current type of e.l.c.b. has certain advantages and is preferred though more expensive. On the other hand, the voltage-operated type requires less current to operate the trip coil. Therefore, where the soil has high resistance – usually more than 40 ohms, but in some cases more than 80 ohms – a voltage-operated type must be installed. The electricity board, which should be familiar with the soil conditions and have instruments for testing the effectiveness of earthing, can usually advise on which type to install.

An e.l.c.b. will not, any more than will a fuse, prevent a person from getting a fatal shock if he touches a live conductor when standing on wet ground or in contact with earthed metal. The reason is that it requires much less current to kill than to operate an e.l.c.b. There are however high-sensitivity e.l.c.b.s available which do provide personal protection from electric shock in those circumstances. Residual-current operated e.l.c.b.s. are now being widely used and are a requirement of the current I.E.E. Regulations for all socket outlets which are for use with portable electric tools used outdoors. This therefore includes garages and workshops because socket outlets in these places are most likely to be used for outdoor equipment. The residual-current operated e.l.c.b. (generally called a residual-current circuit breaker or r.c.c.b.) operates on an imbalance of current in the live and neutral conductors of the circuit. It cuts the current off instantly and is available as a plug-in device into which a power tool can be plugged.

3

INSTALLING THE CABLES

The home wiring installation is a network of cables and wires arranged as individual circuits of various current ratings, all originating at the consumer unit or at main switches and fuses.

The consumer unit is therefore the distribution point of the installation. Comparatively large cables feed the consumer unit from the electricity board's supply and smaller cables carry the electric current from the consumer unit to lights, electrical appliances and other current-consuming electrical apparatus (see Fig. 12).

The current ratings of the circuit are determined by the functions of the circuit and the maximum amount of electrical cur-

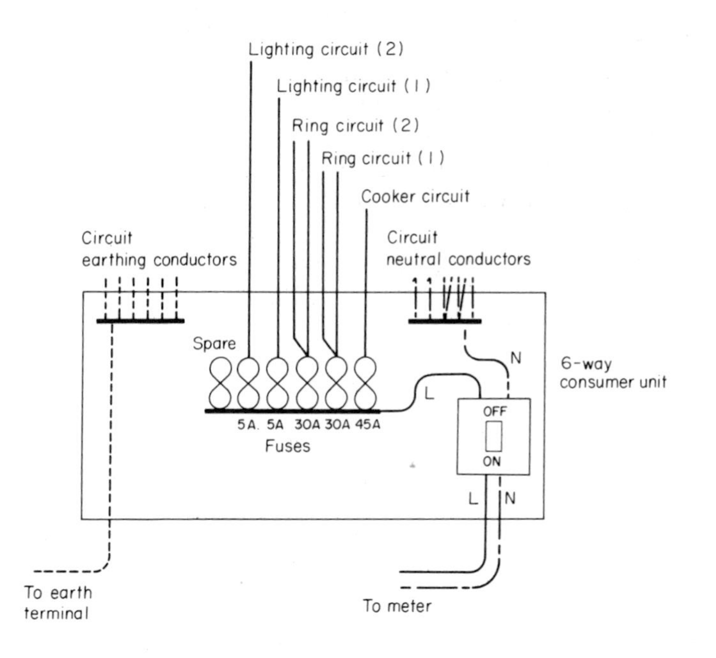

Fig. 12. Connections of circuit cable conductors, mains leads and earthing leads of a typical consumer unit.

rent the circuit is expected to carry. We have already seen that the current ratings of home circuits are standardized at 5A, 15A, 20A, 30A and 45A, these representing the current ratings of fuses or m.c.b.s.

It is in fact a requirement of the regulations (IEE Wiring Regulations) that the current rating of a fuse or m.c.b. must not exceed the current rating of the smallest cable in a circuit, including any flexible cord that forms part of a circuit. Circuits and their current ratings are listed in table 2.

Installing the cables is the major part of home wiring and the important preliminary of any wiring job.

Ideally, all wiring should be installed before any switches, ceiling roses or other accessories are connected and fixed. When wiring a house during the course of its construction, this is called 'carcasing', but most wiring carried out by the householder is in a completed house. However, whether it is a complete house to be wired or rewired or only one circuit or an extension to a circuit, the technique is much the same.

Table 2 House wiring cables

Size (sq.mm)	Current rating (amps)	Circuit	Circuit fuse (amps)
1.0	14	Single light	5
1.5	18	Lighting – loop-in and radial	5
2.5	24	Ring circuits. Radial circuits for up to 20m^2 floor area. Storage heaters	30 20 for radial and heater circuits
4	32	Radial circuits for up to 50m^2 floor area. Also small cooker	30
6	40	Large cooker	45
10	53	Meter tails specified by	
16	70	local Electricity Board	

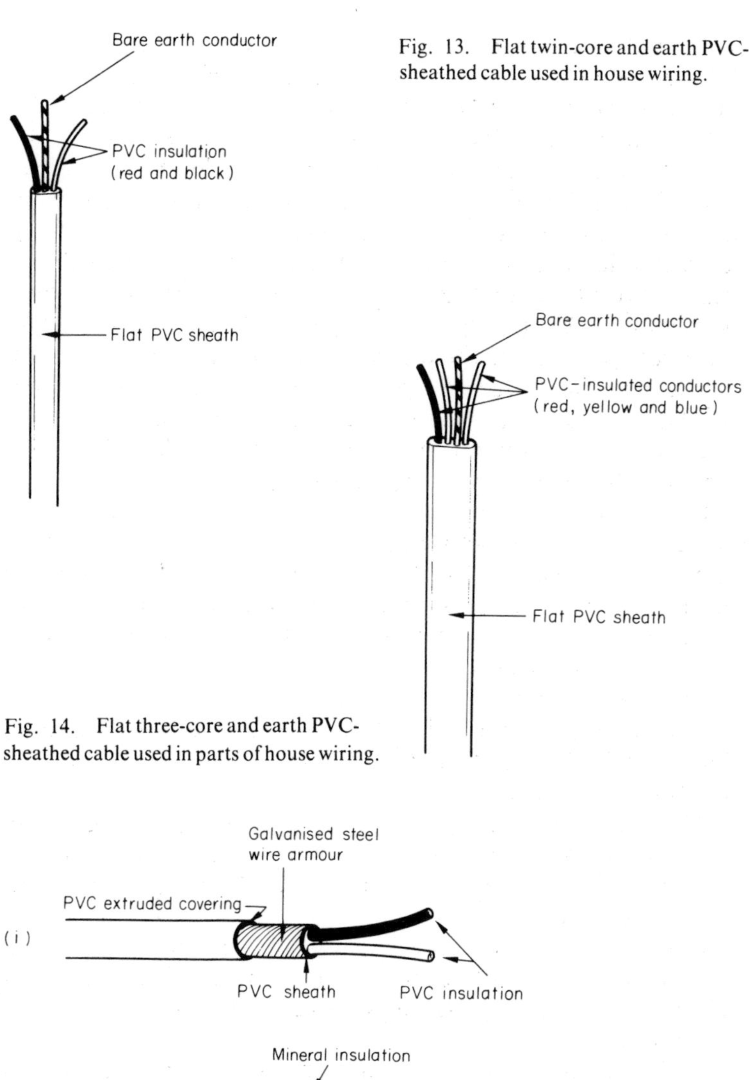

Fig. 13. Flat twin-core and earth PVC-sheathed cable used in house wiring.

Fig. 14. Flat three-core and earth PVC-sheathed cable used in parts of house wiring.

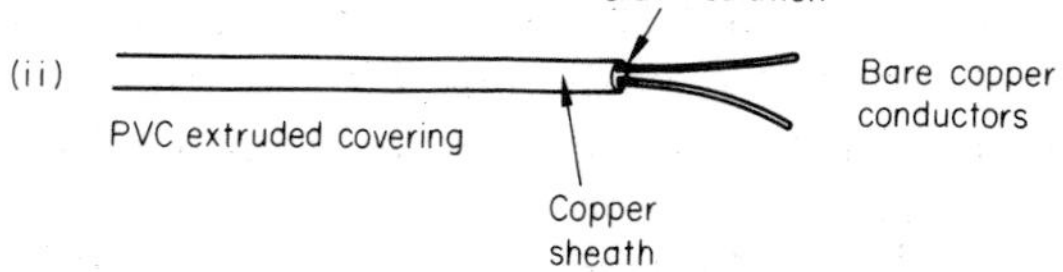

Fig. 15. a) Armoured PVC/PVC cable used outdoors, especially for laying underground; b) mineral-insulated copper-sheathed (m.i.c.c.) cable with PVC covering for outdoor situations including underground.

TYPES OF CABLES

House wiring cables are made in various types and sizes. For interior wiring, the type of cable most commonly used is flat twin-core and earth PVC-sheathed. This consists of two PVC-insulated conductors, one having red insulation, the other black insulation. These are the circuit current-carrying conductors (Figs. 13, 14 & 15). Running between these two insulated conductors is an uninsulated conductor which is the earth continuity conductor (e.c.c.). This carries current only when there is a fault in the circuit, its function being to operate the fuse or m.c.b. The three conductors are enclosed in a tough PVC sheathing, usually grey in colour although also available in white for use in surface wiring. The conductors (wires) are usually of copper but an alternative is copper-covered aluminium, which is lighter and a little cheaper.

Cables used in home installations are in five principal sizes (see table 2). The three smaller sizes have single-strand (solid) conductors; the larger sizes are seven-strand conductors. Some parts of an installation – notably two-way and three-way (intermediate) switching circuits – use flat three-core and earth PVC-sheathed cable.

The three current-carrying conductors are coloured red, yellow and blue respectively; the e.c.c. is uninsulated. The core colours have no significance in home wiring but are used for identification purposes.

The cables connecting the consumer unit to the electricity board's meter are single-core, PVC-insulated and sheathed. The insulation of one (the live) is coloured red, that of the neutral is coloured black.

Cable used for earthing switch gear, for bonding to earth extraneous metalwork, and as an e.c.c. run independently of the circuit cable is single-core green yellow PVC-insulated (non-sheathed) cable.

PVC-sheathed cable, which has an indefinitely long life, has the advantage that it can be run in practically any situation within the house without the need for any further protection from mechanical damage. Only cables run on the surfaces of walls, where there is a risk of damage, need to be enclosed in conduit or have other protection.

Cables run entirely in conduit or enclosed in trunking can be non-sheathed. These are termed conduit or trunking installations, but they are rarely installed in homes today and are not considered suitable for do-it-yourself work.

Outdoor cables. Special cables are usually required for installations out of doors. There are two principal cables used in these situations: one is mineral-insulated, copper-clad (m.i.c.c.), having an extruded covering of PVC to protect the copper sheath from corrosion; the other is armoured PVC/PVC cable which also has an extruded covering of PVC. An alternative is ordinary PVC-sheathed house wiring cable run in heavy-gauge galvanized steel conduit or an approved plastic conduit of the rigid type. (See 'Outdoor Wiring', p. 115.)

TOOLS

Most of the tools required are those the average householder has in his tool kit, though some new ones may be needed.

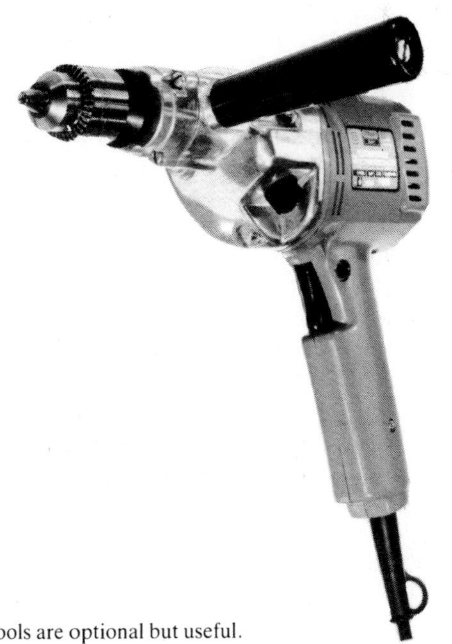

Fig. 16. Power tools are optional but useful.

Essential tools are: tenon saw, padsaw, hacksaw; claw hammer, heavy hammer; nail punch, 150mm cold chisel, 300mm cold chisel, bolster chisel, wood chisels (various); screwdrivers (various); steel tape; plumb line; ratchet brace and various diameter bits. A power tool is optional (see Fig. 16).

PLANNING THE WORK.

First decide on the number of lights, switches, other outlets and any fixed appliances. Decide on the exact position of each and mark the positions on walls and ceiling with chalk. If rewiring, some lights and switches will be in existing positions so these should be noted.

Next, roughly plan the routes of the cables. Routes should be as short as possible but as obstacles may be encountered, for example under floors, some changes in routes may be necessary. Cable is comparatively cheap, so when planning routes in the roof space don't take short cuts.

Cables are run in the following positions:

(i) in the roof space between and across joists;
(ii) beneath floorboards in the void between floorboards and ceilings;
(iii) beneath floorboards of suspended floors on the ground floor where the cable may rest on the concrete of the foundations;
(iv) on the surface of ceilings where there is no access above the ceiling;
(v) on the surface of walls;
(vi) buried in the plaster of walls;
(vii) in existing conduits buried in the wall.

Most of the wiring will run under the floor. This involves moving heavy furniture and lifting floor coverings. Where this is likely to be difficult it may be better to run the cable along a more circuitous route.

RAISING FLOORBOARDS

Having decided on the approximate routes of the cables to be run under the floor select the boards to be raised.

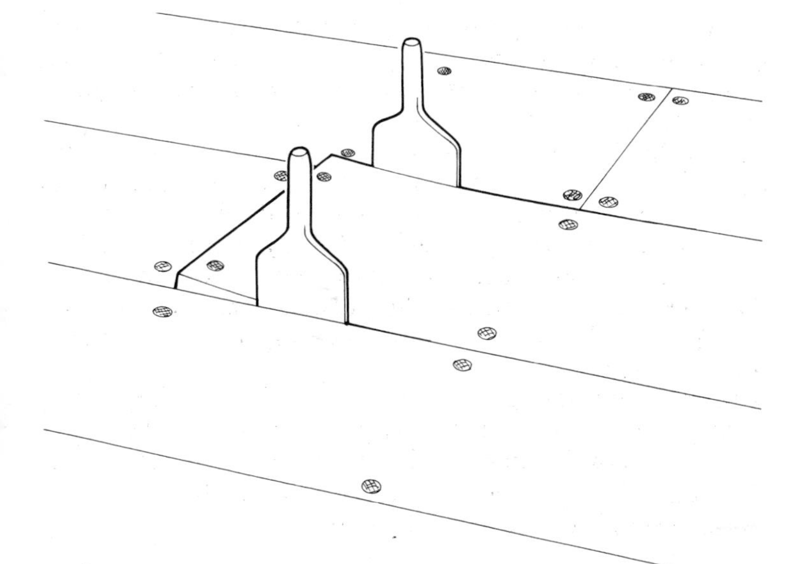

Fig. 17. Two bolster chisels are used to start lifting a floorboard at a butt joint.

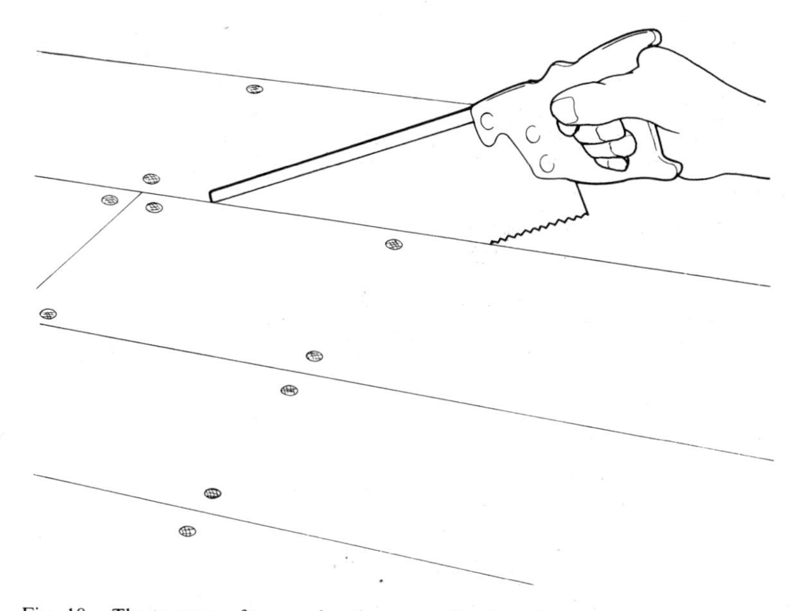

Fig. 18. The tongues of tongued-and-grooved floorboards must be cut off before the board can be raised.

The first choice will be a board that has obviously been raised before and is only lightly nailed or screwed. The next choice will be a board that has a joint somewhere along its length which is the point where you can most easily start raising it (Fig. 17). If a board must be raised that has no joint throughout its length you will need to cut it at a joist before you can raise it, using the method described below.

Lift boards sparingly. That is, lift only those essential for running the cables. Much of the wiring can be done by 'fishing' the cables out from under the boards but where the cable is to cross joists under boards a board extending the whole length of the cable run must be raised.

In the conventional two-story house the principal or key board to raise on the first floor is one running the full length of the centre of the landing and through the room at each end. In this way you can gain access to most of the lighting points and switches simply by fishing out the cables.

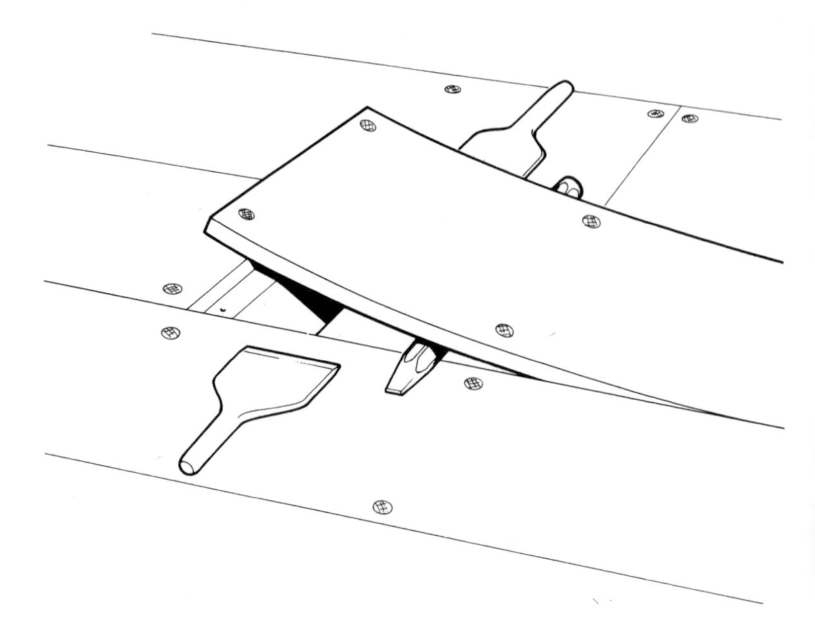

Fig. 19. Cold chisel under a partly raised floorboard. This is pushed along progressively as the board is lifted to 'spring' the board out.

Tongue and grooved boards. Before a tongue-and-groove board can be raised it is necessary to run a hand saw down each side of the board to remove the tongue interlocking the board with the adjacent boards (Fig. 18). When cutting the tongue take care not to cut cables below the board.

Lifting a previously relaid floorboard. If refixed by screws, first remove the screws and lift the board using a bolster chisel. If nailed, the board can be lifted in the same manner but you will need to use extra exertion as well as a second bolster chisel or a 12in cold chisel (Fig. 19).

Lifting a jointed board. Insert a bolster chisel at each side of the board near the joint, having first punched through the nails securing the board to the first two joists. Prize up the end of the board using the bolster chisel and insert a 300mm cold chisel under the floorboard with its ends resting on the two adjacent floorboards.

Raise the end of the board higher and push the cold chisel further along the board. Continue this process and then, by

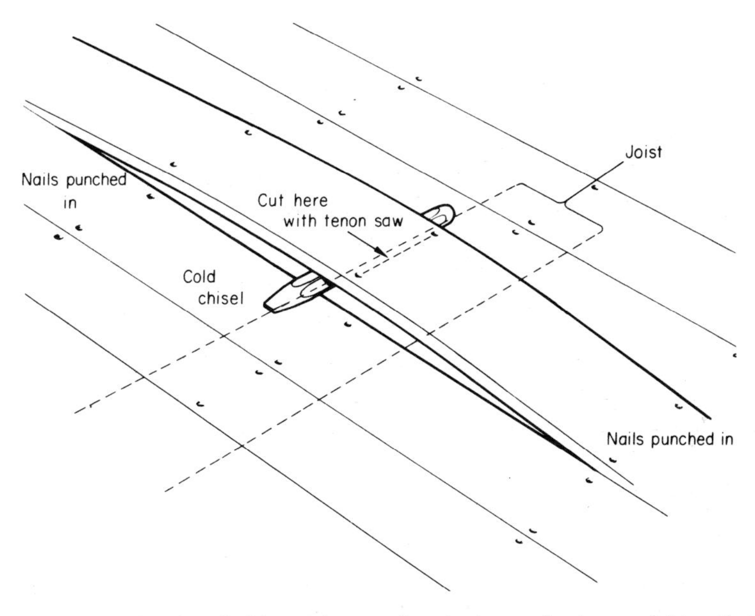

Fig. 20. A floorboard with no joint must be prized up and cut over a joist so that either one or both sections can be raised.

alternately lifting the board and pressing it down, using the cold chisel as a fulcrum, you will prize out the nails and the board will be free. The end may have to be wrenched from under the skirting similarly to the extraction of a tooth but take care not to damage the skirting board.

Lifting a jointless board. Choose a position at a joist where the board can be conveniently cut (Fig. 20). This will be at a point where you require access only under a section. Using the nail punch, punch through the nails at this joist and at two joists on each side. At the first joist chosen (centre joists without nails now securing it), insert a bolster chisel under each side and prize the board up sufficiently to insert a 300mm cold chisel underneath with its ends resting on the adjacent boards. Now, using the tenon saw, cut the board at the point coinciding with the centre of the joist so that when relaid both sections will be supported by the joist.

If only one section of the cut board is to be raised place the cold chisel under this in the first instance before cutting the board. Raise the end as for the jointed board and progress with the cold chisel in the same manner until the board is free. If the other section is to be raised also deal with this similarly; if not you can nail it in position.

REPLACING FLOORBOARDS

It is as well not to refix any boards until the wiring is complete but to lay them in position for safety so you do not step between joists and put your foot through the ceiling.

First remove all nails from the boards and any nails that remained in the joists. Check that the joists are clear of wood chippings and other foreign matter. Check also that no tools or other objects have fallen into the void.

Make sure that each board is laid 'the right way round'. If there are no distinctive markings you should pencil lines across boards before lifting a board, so that you can match up the pencilled lines when relaying.

When you finally relay boards after wiring it will be necessary to cut a slot in the end of any board where a cable passes up behind the skirting board to a socket outlet. Where existing cables are laid in slots cut in the tops of joists, check that no cable is out of a slot, which would cause the board to ride.

Cut traps (short lengths of floorboard) above a lighting point and secure these by screws so that they can be readily raised if necessary.

Use 50mm oval nails to fix the floorboards but make certain they do not pierce cable or pipes. If, as is correct, these are at the centre of floorboards and the boards are traditionally nailed near the edges, there will be no such risk. But when the house was wired or piped during its construction the position of the floorboards will not have been known and some cables and pipes may well be off-centre.

DRILLING JOISTS

Cables crossing joists beneath floorboards must be threaded through holes drilled in the joists at a height of not less than 50mm below the top of joists (Fig. 21). For this use a ratchet brace and a 19mm bit, or a power drill at a low speed.

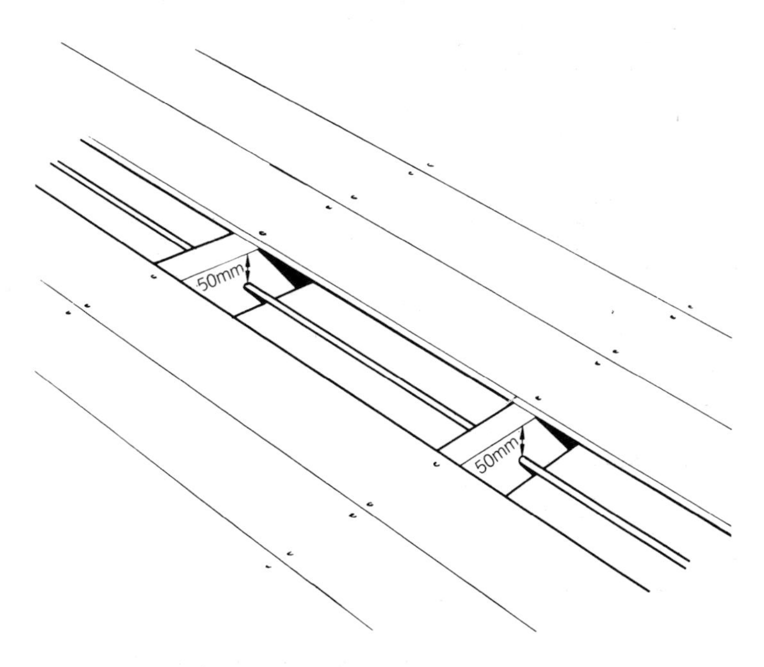

Fig. 21. Sheathed cable beneath floorboards is threaded through holes drilled in the joists at least 50mm below the tops of these joists.

SECURING CABLES

Cables run under floorboards do not have to be fixed but may rest on the structure if they are unlikely to be disturbed. In an unboarded roof space the cables can be fixed to joists as necessary to secure them, but if the space is used for storage or if there is access to the cold water cistern it may be necessary to run some of the cables through holes drilled in the joists, as under floorboards.

Cables run on surfaces. Cables run on the surface of walls are fixed by clips spaced not more than 225mm apart in horizontal runs, and not more than 375mm apart in vertical runs. For neatness, and to prevent sagging, preferred spacings where the cables are exposed to view are 150mm for horizontal and 225mm for vertical runs (Fig. 22). Sagging can be eliminated entirely by smearing the back of the cable with contact adhesive.

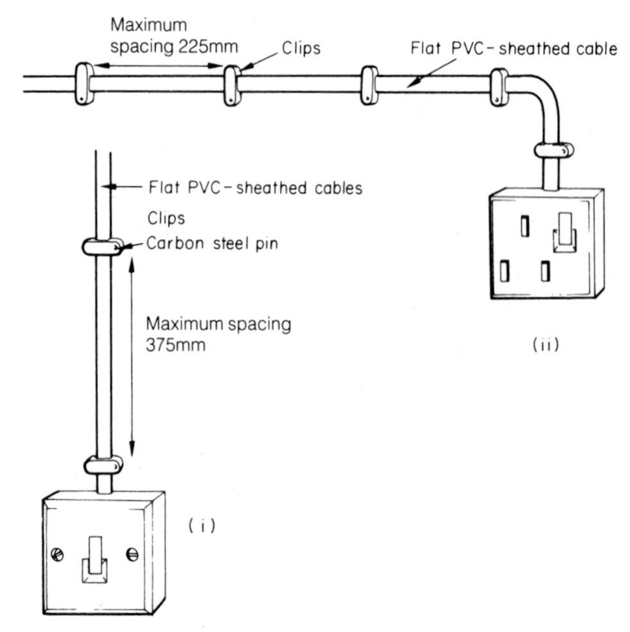

Fig. 22. (i) Fixing sheathed cable vertically down the surface of a wall.

(ii) Fixing sheathed cable horizontally along the surface of a wall.

For fixing cables you can use plastic clips that are secured by carbon steel pins that come attached to them. These clips are available in various sizes to suit the cable. There is no need to plug the wall – just hammer the pins in. To get a straight line you can either pencil a line against a yardstick or snap a chalked plumb line.

Alternatively, cable may be enclosed in wood capping or in plastic oval conduit, or in plastic mini-trunking. Where a cable is to be run across a ceiling to a lighting point because there is no access above, mini-trunking or plastic oval conduit usually necessary to prevent sagging.

Sinking cables in walls. At drops to switches and to wall lights cables are best buried in the plaster(Fig. 23).

To do this, mark two lines down the wall from the ceiling to the switch or wall light using a straight edge or chalked line. Spacing between the lines should be identical to the width of the cable.

Then, using a straight edge, run a Stanley knife down each line to start the chase in the plaster and (where relevant) to cut the wallpaper. Remove the plaster using a sharp bolster chisel and clear the chase with a sharp cold chisel.

Where the plaster is only a thin coating some brickwork will also need chasing out to deepen the channel to allow sufficient plaster filler to secure the cable.

Place the cable in the chase and fill it with a proprietory plaster filler. If it is difficult to retain the cable in the chase smear the back surface with contact adhesive as for surface cables.

Where a wall is not to be redecorated immediately it is better to fix the cable to the surface and sink it later when repapering.

Where there is conduit sunk in the wall down to an existing switch position this may be used for the new cable, but if it is a metal conduit fit a rubber or PVC bush on each end to prevent damage to the cable sheath.

As already mentioned, it is not necessary to enclose sheathed cable in conduit or cover it with capping when sinking it in the wall. The reason for this is that burying the cable gives it protection from external damage. However, neither conduit nor metal channel will prevent a drill or plugging tool from damaging the cable. The solution to this is, when fixing

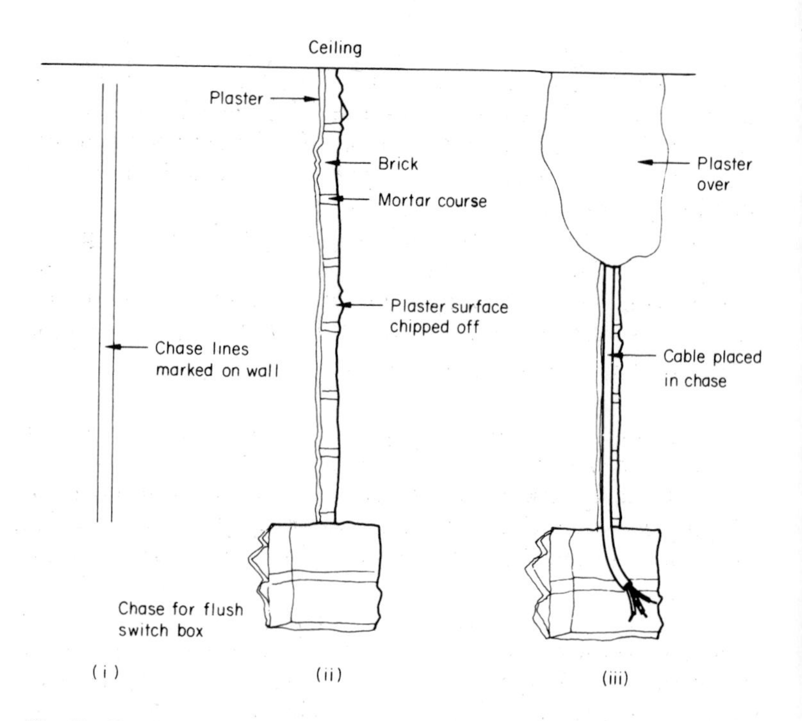

Fig. 23 Burying sheathed cable in the plaster of a wall: (i) chase marked on wall with pencil: (ii) plaster removed; (iii) cable laid in chase and then plastered over. Chase also cut for switch box.

shelves or other objects to avoid obvious cable runs. All cable must be run vertically either up or down from a switch, socket outlet or other accessory. No cable must run horizontally or at an angle. The exceptions are – close to the ceiling where access from above is not possible or along the skirting board either in front or behind it.

4

THE LIGHTING CIRCUIT

Two systems are used for wiring lighting circuits. One is the loop-in system; the other is the joint box system.

The loop-in system is the preferred and more popular method. The joint box method is confined mainly to situations where looping would be difficult; can be a lighting circuit, which is a mixture of both methods.

THE LOOP-IN SYSTEM

The principal electrical accessory of the loop-in system is a loop-in ceiling rose, which serves as a joint box as well as performing its normal function as part of a lighting pendant to which the flex is attached (see Fig. 24). The looping-in of cables is not however confined to ceiling roses. It is done also at the ceiling plates of pendant fittings, close-ceiling lighting fittings and batten lampholders.

With the loop-in system a twin-core and earth PVC-sheathed cable consisting of live and neutral conductors as well as an earth wire runs from the lighting circuit fuse way in the consumer unit to each of the lighting points of the circuit.

The cable runs to the first lighting point, which is the one nearest to the consumer unit, is looped in and out of this point, runs to the next, is looped in and out of it, then to the next and so on until it reaches the last on the circuit, where it is terminated. This means that every lighting point has a live and a neutral conductor which come direct from the circuit fuse.

As each lighting point is normally required to be separately switched, a length of the same type and size of cable is run from each lighting point to its switch. The switch is usually fixed to the wall in the same room but is sometimes a cord-operated ceiling switch. Additional wiring is required where a light is to be switched from two or more positions.

This method of wiring means that at each lighting point there are three twin-core and earth PVC-sheathed cables: the incoming feed cable, the outgoing feed cable to the next light, and the cable to the switch. At the last lighting point on the cir-

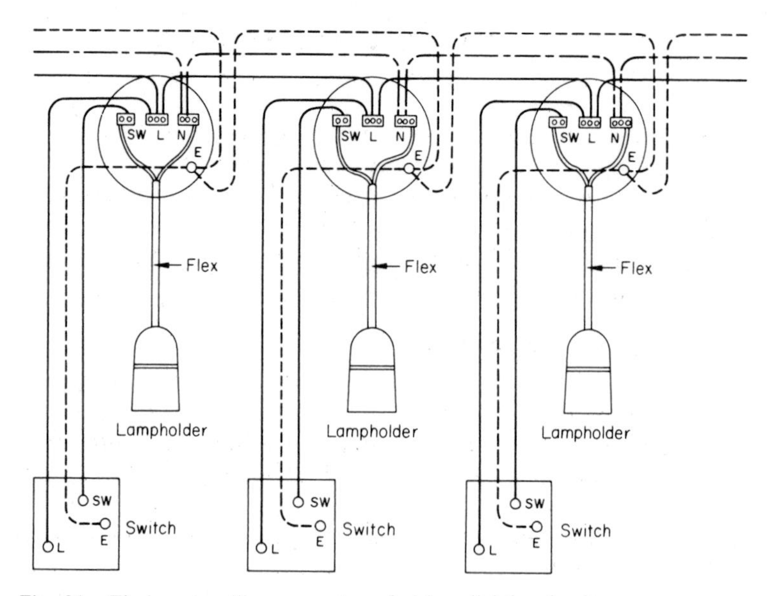

Fig. 24. The loop-in ceiling rose system of wiring a lighting circuit.

cuit there are two cables only since there is no outgoing feed cable to another point.

The red core of each cable is the live conductor throughout the circuit. The black conductor of the feed cables is the neutral, but the black conductor of the cable between the light and the switch is a switch return cable and is therefore a 'live' cable. This has to be identified by enclosing the ends with red PVC sleeving, but sometimes you can buy a twin cable having two red insulated conductors.

The connections at a ceiling rose are shown in (Fig. 24) where it will be seen that the twin flex of the pendant is connected to the neutral and switch wire terminals. The live loop in conductors is jointed at a terminal that does not carry a flex conductor. The earth wires (e.c.c.s) are jointed at an earth terminal. Where lighting fittings other than plain pendants with ceiling roses are installed, the flex connections are made at a flex connector housed in a ceiling box. The live looping conductors are jointed also in a connector, and the e.c.c. goes to an earth terminal. If the fitting is of metal this must also be connected to the earth terminal by a length of PVC insulated earth wire. A similar arrangement is made for close-ceiling fittings, depending on the type, and also for wall lights, but these do not usually have loop-in cables (see 'Wall lights', p. 55).

THE JOINT BOX SYSTEM

When joint boxes are used instead of loop-in ceiling roses, the twin-core and earth PVC-sheathed cable runs from the fuse way in the consumer unit to a series of joint boxes and not direct to the lighting points (Fig. 25). These joint boxes have four terminals and are fixed to timber secured between two joists about equidistant from a light and its switch. This means that a separate joint box is needed for each light and switch.

From each joint box one twin-core and earth PVC-sheathed cable is run to the respective light and another to the switch. The connections at the joint box are shown in (Fig. 26). At the ceiling rose or other lighting fitting the two current-carrying insulated conductors are connected to the two terminals carrying the flex. The single earth conductor is connected to the earth terminal. The connections at switches are the same as for the loop-in system.

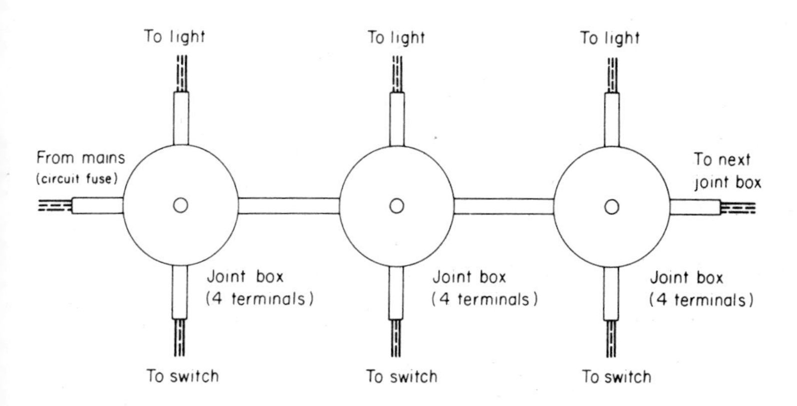

Fig. 25. Layout of the joint box system of wiring a lighting circuit.

TWO-WAY SWITCHING CIRCUITS

A two-way switching arrangement enables a light to be switched on and off from two positions. Examples are: a light on the landing switched from both landing and hall; a bedroom light switched at the door and at the bedhead; a kitchen light switch at the access door from the hall and also at the back door. Each of the switches of a two-way switching circuit is similar to an on/off switch but has three terminals instead of the usual two.

There are various ways of wiring and connecting two-way switching circuits and switches, but when wiring a circuit with twin-core and earth PVC-sheathed cable it is best to wire the first of the two switches as though for one-way and from this switch to run another cable to the second two-way switch. The cable linking the two-two-way switches is three-core and earth PVC-sheathed cable already referred to. There will be two cables running down to one of the switches, a two-core and a three-core. The connections at each of the switches is shown in Fig. 27.

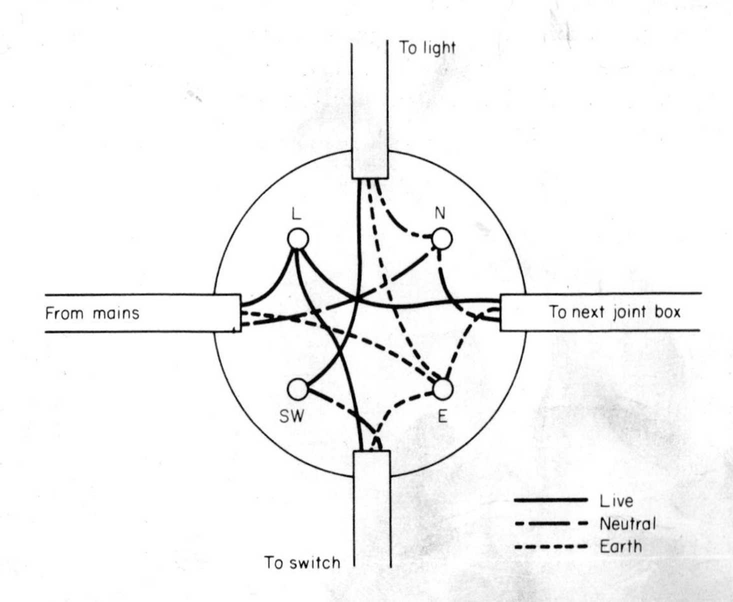

Fig. 26. Cable connections within a joint box of a lighting circuit.

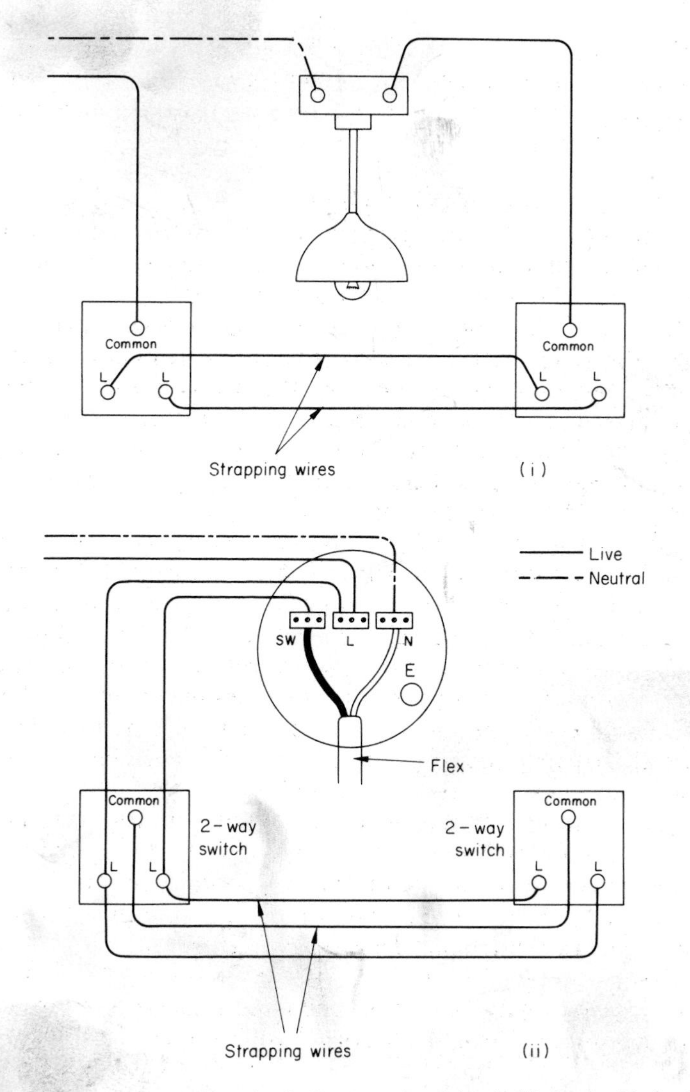

Fig. 27. Two-way switching circuit: (i) conventional circuit; (ii) circuit with sheathed cable system.

When one switch is a cord-operated ceiling switch the three-core and earth cable is run to this, but to save running two cables down to the wall switch, treat the ceiling switch as the first switch of the two-way arrangement. This will mean running the twin core cable to the ceiling switch as though for one-way working and from this running the three-core cable to the wall switch.

INTERMEDIATE SWITCHING

An intermediate switching circuit enables any one light or cluster of lights, or a number of lights in a stairway, hall or landing, to be controlled from three (or more) positions.

The circuit is basically the same as for two-way switching but with the addition of a special switch, (or switches) inserted in intermediate positions between the two two-way switches. For control at three positions one intermediate switch is required in addition to the two two-way switches. For more than three switch positions an additional intermediate switch is required for each additional switching position.

When wiring the switching circuit, first decide on the positions of the switches and wire the first as for on/off or one-way working, which means running a twin-core and earth sheathed cable from the light or joint box. Then from this first switch of the system run a three-core and earth PVC-sheathed cable to each in turn of the other switches of the intermediate circuit. At each intermediate switch there will be two three-core and earth cables, but there will be only one cable at the last switch, which is a two-way switch.

The connections at the switches are shown in (Fig. 28) where it will be noted that only two of the three conductors of each cable are connected to the intermediate switch terminals (one to each of the four terminals), the third acting as a through conductor running from the 'common' terminal of one of the two two-way switches to the 'common' terminal of the other. This common wire can be the red conductor, the yellow and blue conductors being used for what is termed the strapping wires.

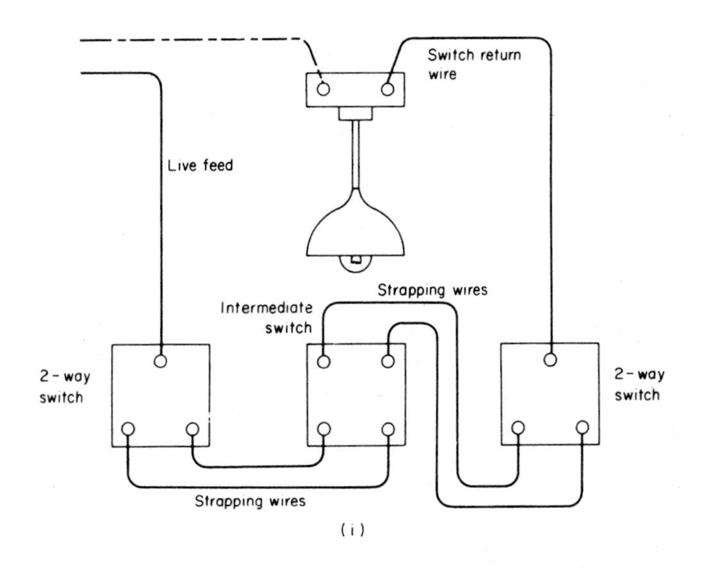

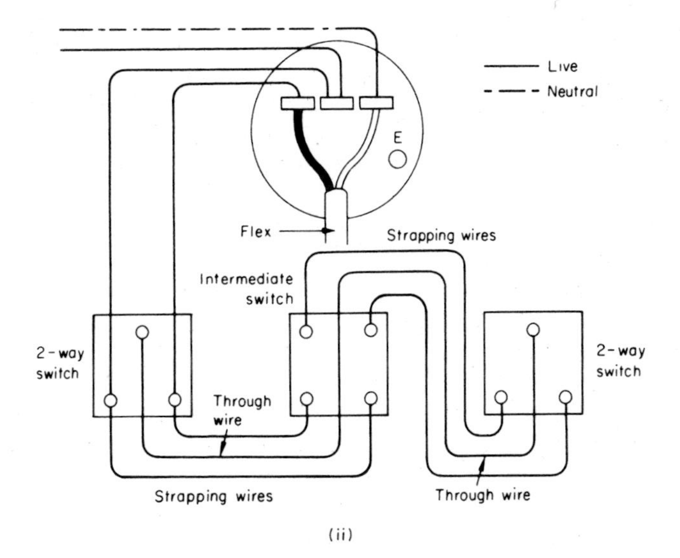

Fig. 28. Intermediate switching circuit where a light is controlled from three or more positions: (i) conventional circuit; (ii) circuit normally used with a sheathed cable system.

5

LIGHTING CIRCUITS ACCESSORIES

The principal wiring accessories used in lighting circuits are ceiling roses, batten lampholders and other lighting fittings, fixing plates or boxes, joint boxes and switches.

Ceiling roses, batten lampholders and most close-ceiling fittings are fixed direct to the ceiling. So also are cord-operated ceiling switches, except for some patterns that have an open back and need to be mounted on a pattress block or box.

Special pendant fittings and most wall lights have to be mounted on a backing box which contains the flex connector and the unsheathed ends of wires.

CEILING ROSES

First make certain that there is adequate backing material to support the ceiling rose. Where possible secure it to a joist. If however the light is positioned between joists fit a piece of timber 75 × 19mm between the two joists immediately above the ceiling, after first drilling a 19mm diameter hole in the timber for the circuit cables (Fig. 29). The hole is drilled to coincide with the hole pierced in the ceiling.

Never rely on laths of a lath and plaster ceiling to support a ceiling rose.

The modern ceiling rose with its enclosed base (Fig. 30) is fixed direct to the ceiling without the need for a plastic pattress. First connect the flex to the ceiling rose and at the same time, if you have decided on the length of flex. fit the lampholder before fixing the ceiling rose.

Now knock out the rectangle of thin plastic in the base of the ceiling rose to take the circuit cables. Thread in the cables and fix the ceiling rose using no. 8 wood screws of adequate length. Strip about 75mm of sheathing from the ends of the cables. Push any surplus back into the ceiling allowing about 12mm of sheathing within the ceiling rose. Strip about 12mm of insulation from the end of each conductor.

Connect the red insulated conductors to the live terminal of the ceiling rose. Sort out the black switch return wire con-

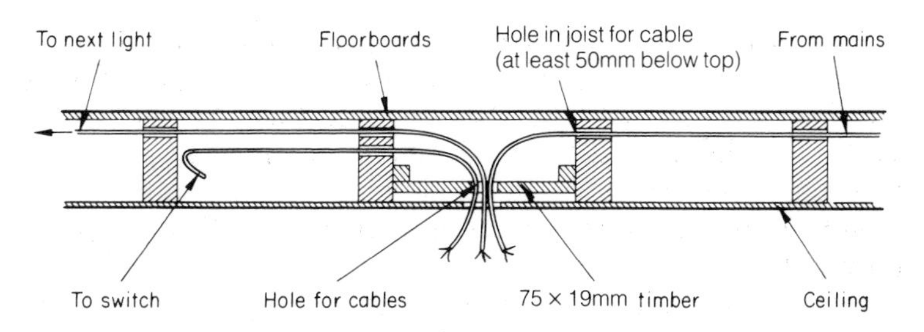

Fig. 29. Timber secured between two joists for fixing ceiling roses and other lighting fittings, including cord-operated ceiling switches.

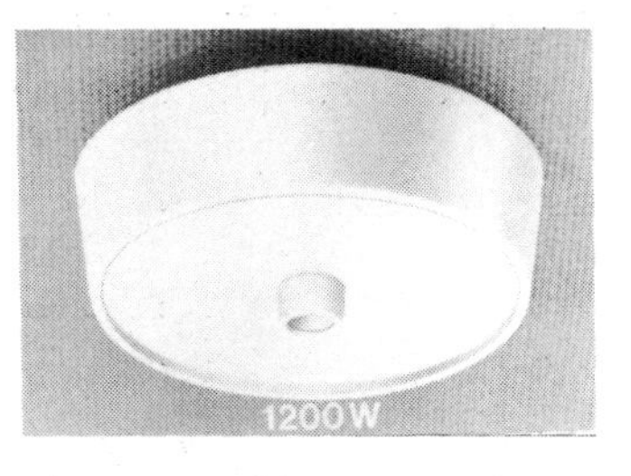

Fig. 30. Modern ceiling roses and lampholders.

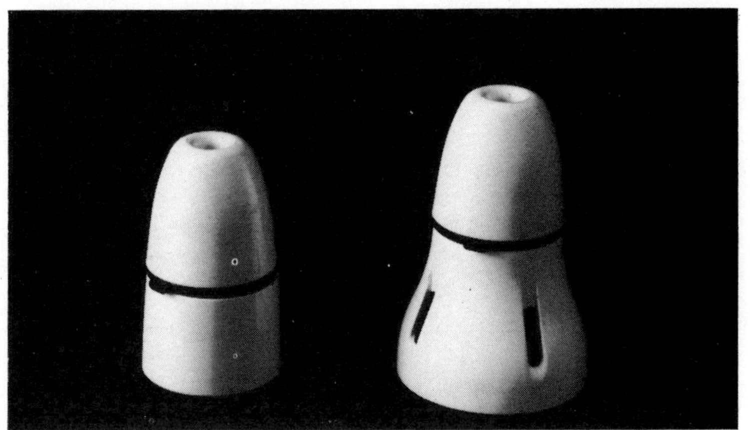

ductor, place red sleeving or red adhesive tape over the end of the black insulation and connect this wire to the switch wire terminal (Fig. 24). Connect the remaining black insulated conductors to the neutral terminal. Slip a length of green yellow PVC sleeving over the bare ends of the earth conductors leaving about 12mm bare for connections. Connect these to the earth terminal.

Check that the flex is properly anchored and screw on the ceiling rose cover. Fit the lampholder to the end of the flex if you have not already done so.

BATTEN LAMPHOLDERS

The batten lampholder is really a simple close-ceiling lighting

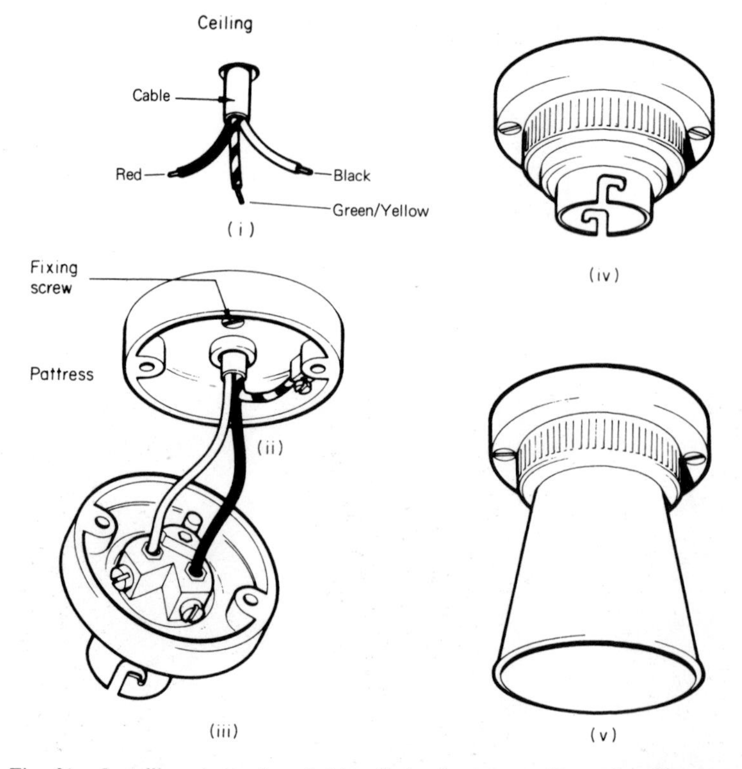

Fig. 31. Installing a batten lampholder. (i) circuit wires at ceiling point; (ii) pattress fixed; (iii) terminal connections; (iv) lampholder fixed to ceiling; (v) deep skirt fitted.

fitting. The modern version is similar to the modern ceiling rose with loop-in wiring facilities and an enclosed back (Fig. 31). It is fixed exactly as a ceiling rose and the connections at the terminals are the same except for the absence of flexible cord.

Screw on the lampholder section and where relevant, such as in the bathroom, fit the deep skirt that shields the brass cap of the bulb from touch when replacing a bulb.

PENDANT LIGHTING FITTINGS

Pendant lighting fittings, usually of metal, and many of the multilight type are bought ready wired and fitted with a flex connector. This connector is connected to the circuit wires at the ceiling point, and the fitting is secured to a joist or to a piece of timber secured between two joists as for a ceiling rose.

There are, however, some differences. A connector is required for the live jointed wires of a loop-in system; the fitting, if of metal, must be connected to an earth wire, and means must be provided for housing the flex connector if the ceiling plate of the fitting has insufficient depth. Also, unless the ceiling plate is of the enclosed type (most unlikely), a pattress

Fig. 32. Rise and fall pendant fitting.

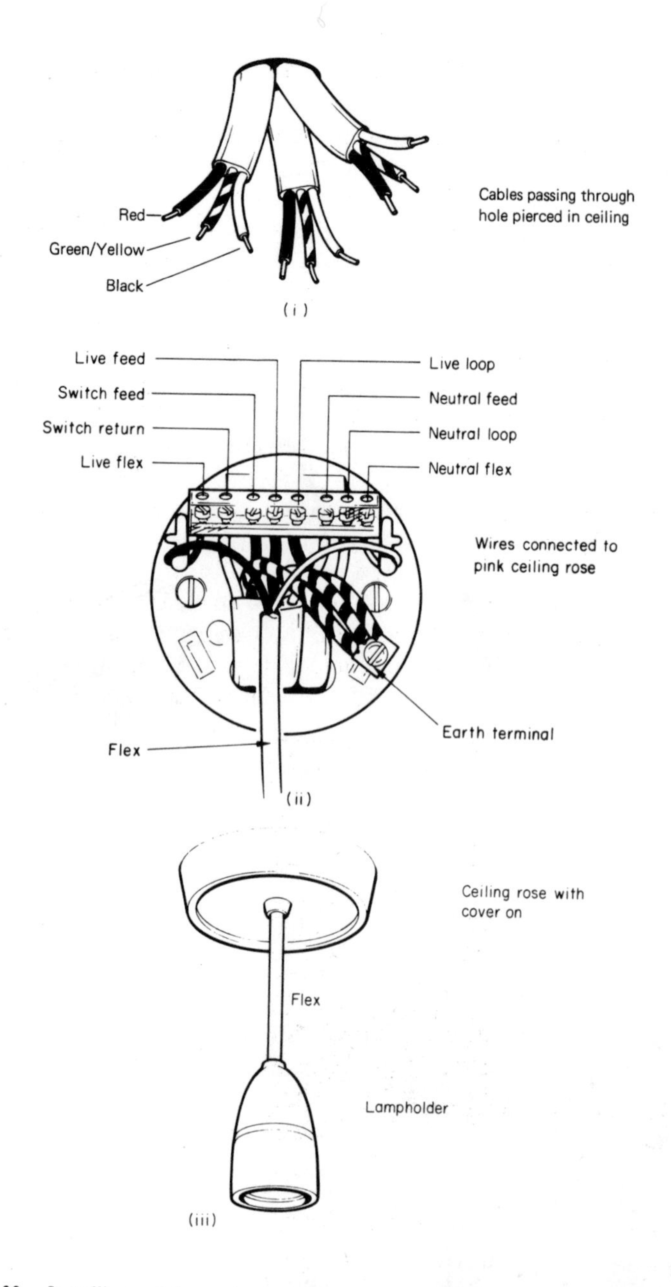

Fig. 33. Installing a plain pendant.

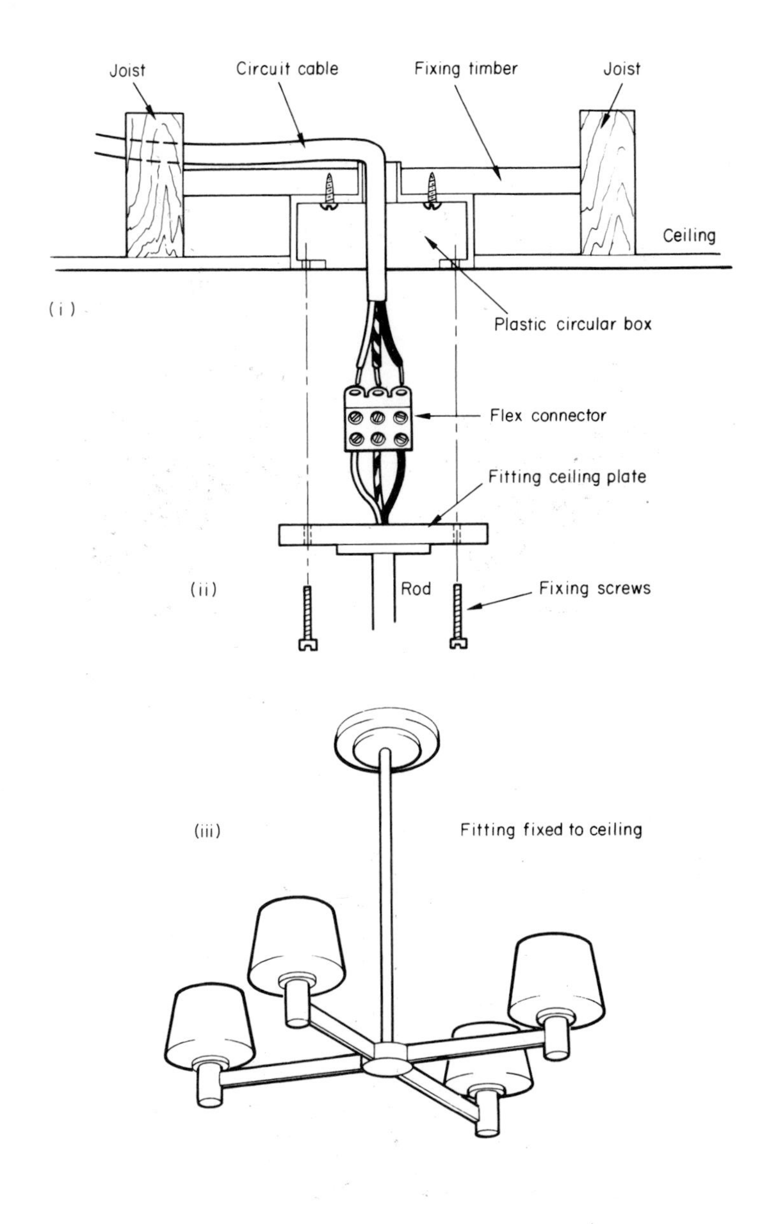

Fig. 34. Installing a multi-light.

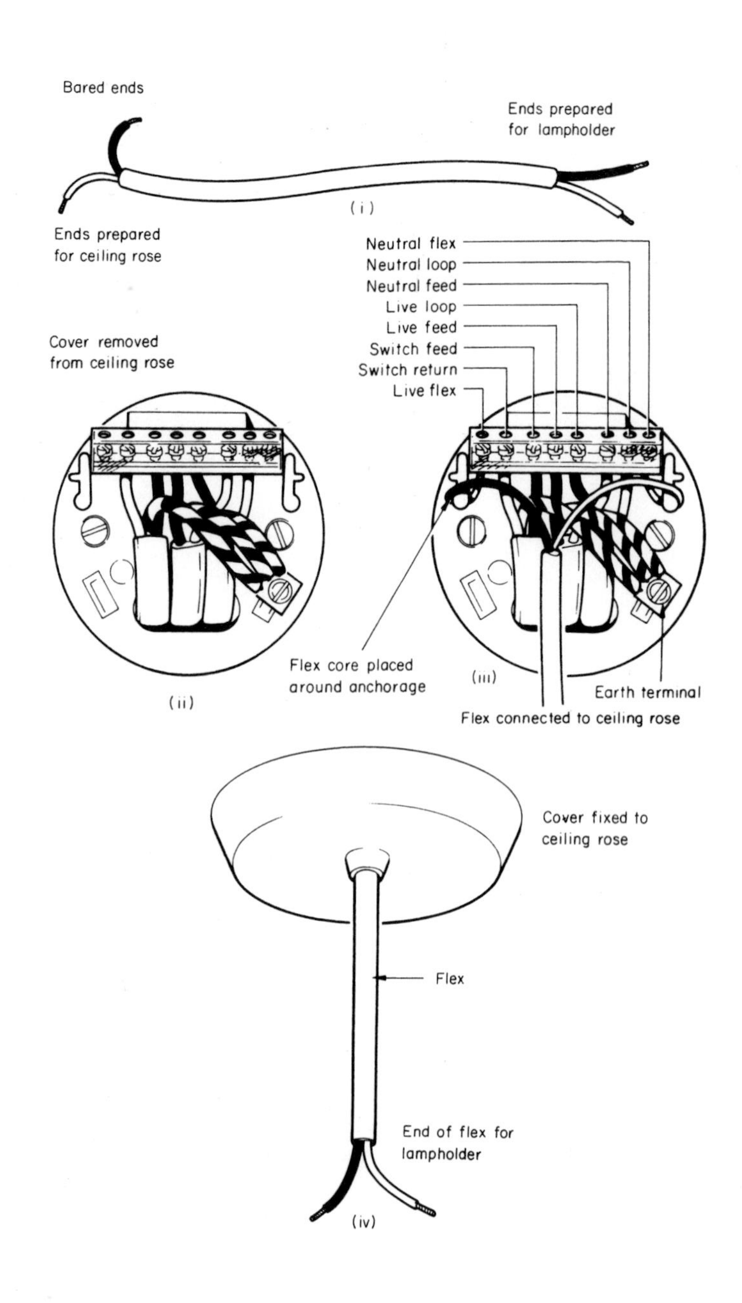

Fig. 35. Fixing flex to a ceiling rose.

block or other means must be found for providing a non-combustible chamber for containing the ends of unsheathed wires.

The solution to all of these problems is to fit a metal or plastic circular box known as a conduit box. This has a rear entry for the circuit cables, two lugs tapped for 2BA or M.4 screws for fixing and supporting the fitting, and an earth terminal for the circuit earth wires (e.c.c.s). From this terminal a short length of green yellow insulated cable runs to the earth terminal of the lighting fitting.

This conduit box can be sunk into the ceiling flush with the plaster or fixed to the surface of the ceiling. Unless the ceiling plate of the lighting fitting fits the box – and most don't – it is necessary to sink the box into the ceiling. To do this cut a circular hole in the ceiling and fix timber between the joists for

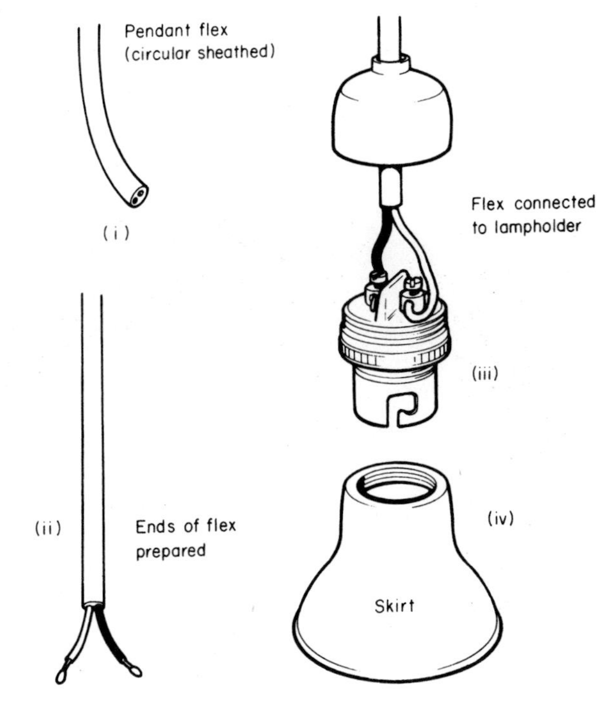

Fig. 36. Fitting flex to lampholder.

securing the box. This will need to be about 45mm above the ceiling to accommodate the box depth. Drill a hole to take the box cable entry lug. Metal boxes are preferable but plastic boxes are usually easier to handle. For heavy lighting fittings if supported by the box lugs always use a metal box. The box lugs have fixing centres of 50mm as have many fitting ceiling plates. Where a ceiling plate has other fixing centres it is necessary to fix a hardwood block to the box and fix the fitting to the block (See Figs. 32–36).

CLOSE-CEILING LIGHTING FITTINGS

The simplest close-ceiling fitting is the batten lampholder already dealt with (Fig. 31). Others include a whole range of differing styles and types including linear and circular fluorescent fittings. Fixing close-ceiling lighting fittings is usually fairly simple. All have an entry hole for the circuit cables in the base of the fitting and most have a terminal block for the two circuit conductors (neutral and switch wire) and an earth terminal. With the loop-in system of wiring the live loop-in wires have to be terminated at an insulated connector housed in the

Fig. 37. Close-mounted ceiling fitting.

fitting. For most fittings the PVC-insulated and sheathed cable run straight to the terminals of the fitting. But where the excessive temperatures are likely to develop, usually in small enclosed fittings having high-wattage bulbs, the circuit wires must be terminated at a joint box fixed between joists above the ceiling. From this, heat-resisting cables are run to the fitting terminals. The live loop-in wires will terminate at the joint box.

Close-ceiling fittings can usually be secured to one or more joists; otherwise timber must be fixed between joists as for ceiling roses (Fig. 29).

FLUORESCENT FITTINGS

No special wiring is required for fluorescent lighting: a fitting is connected as for any close-ceiling fitting (Fig. 39).

Linear fittings can be secured to joists but where the fixing holes do not coincide with the joists more holes must be drilled. Cable entry holes in linear fittings vary, so it is wise to check this before wiring the lighting point. When a fluorescent fitting is bought to replace a filament fitting and the position of the point is fixed it may be necessary to cut a new cable entry hole in the fitting.

Metal fittings must be connected to an earth wire. With fittings of the quick-start type earthing is essential since it is needed to assist in the starting of the lamp when switched on. See Fig. 40.

WALL LIGHTS

Wall lights also cover a very wide range of types and styles (Fig 41). Since it is the backplate that secures the fitting to the wall, the shape and size of this piece is important.

Many backplates are narrow and shallow, with nothing to house the flex connector and ends of the wires and cables. The solution here is to fit a box into the wall flush with the plaster to accomodate the cables and connector. One narrow box that will be covered by most wall-fitting backplates is a metal flush box designed for architrave switches (Fig. 42). To fix this box, first cut out a chase in the wall at the position of the fitting. Then insert a PVC grommet in the knockout hole on the top edge of the box, thread in the sheathed circuit cable and fix the

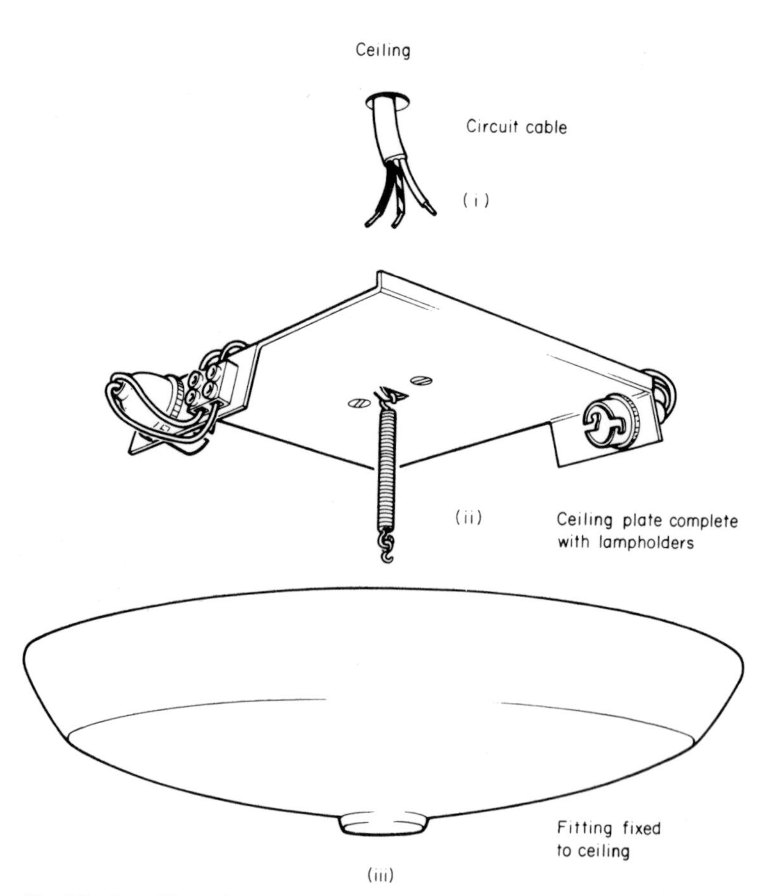

Fig. 38. Installing a close-mounted ceiling fitting.

box with wood screws in plugged holes. Strip off the sheathing, leaving about 12mm within the box, and strip off 12mm of insulation from the ends of the two current-carrying conductors. Insert the bared ends into the flex connector, making sure that the red conductor, which is live, is inserted in the terminal containing the brown flex lead. This is especially important with a wall light that has an integral switch. Slip a length of green yellow PVC sleeving over the bare end of the earth wire before connecting this wire to the box earth terminal. If the fitting is metal, connect an earth wire to its earth terminal.

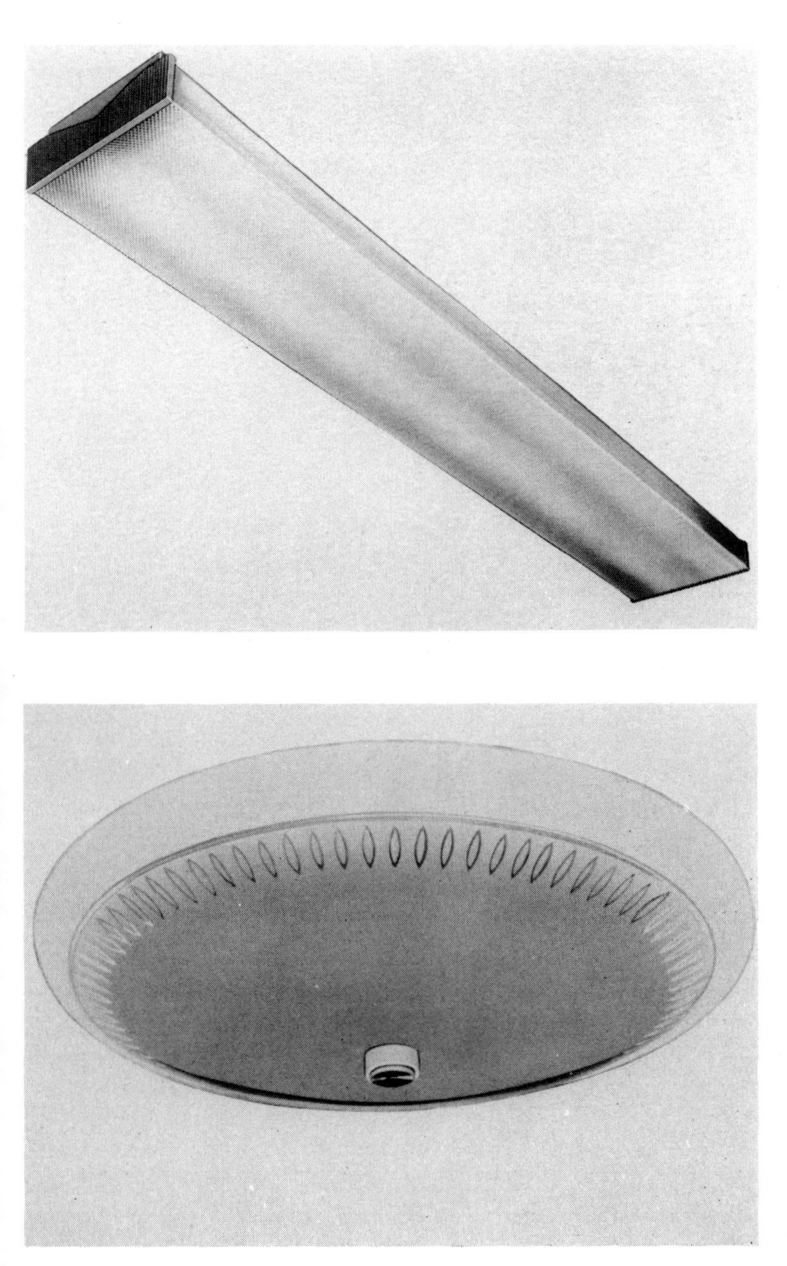

Fig. 39. (a) and (b) Fluorescent lighting fittings.

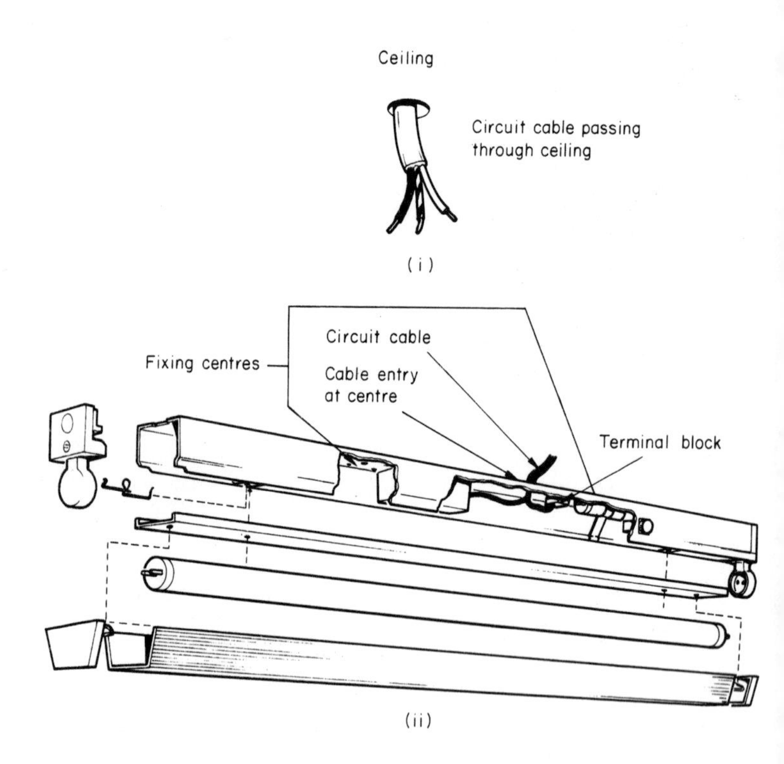

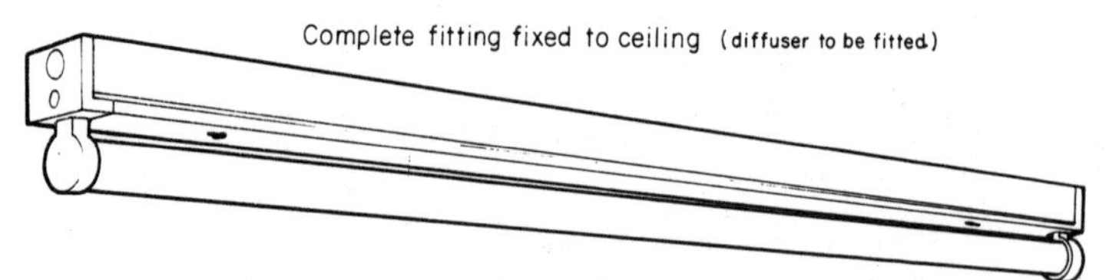

Fig. 40. Installing a fluorescent fitting.

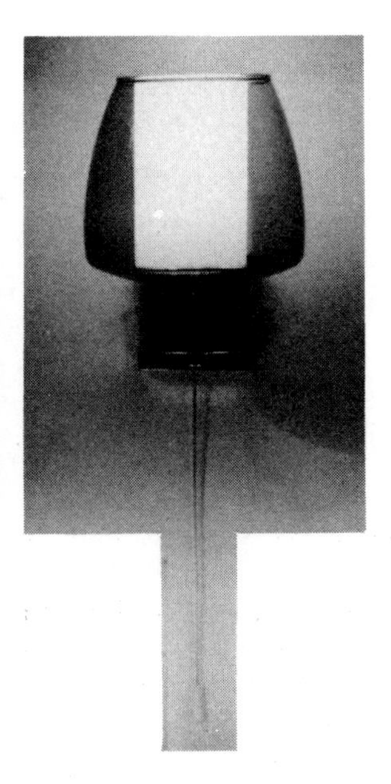

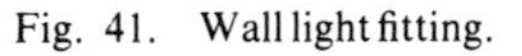

Fig. 41. Wall light fitting.

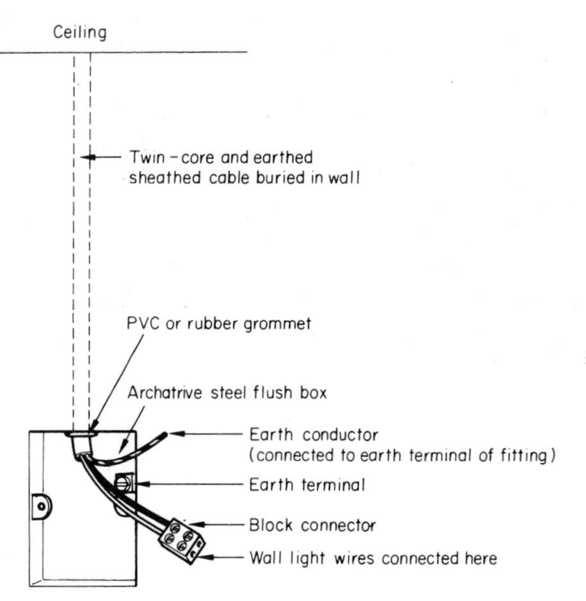

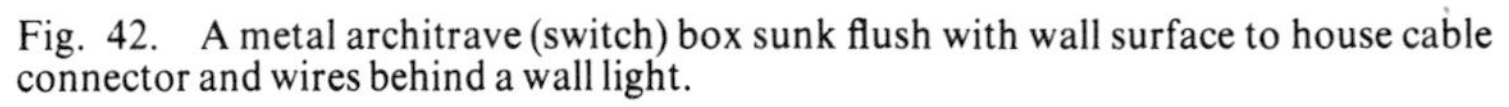

Fig. 42. A metal architrave (switch) box sunk flush with wall surface to house cable connector and wires behind a wall light.

The fitting itself is screwed direct to the wall using wood screws since the lugs of the box are designed for a switch, not for a wall light.

Some wall fittings have circular backplates with 50mm centre fixing holes. With these use the standard circular conduit box already described. The wall light is fixed to the lugs of this box using 2BA or M.4 metric screws.

SPOTLIGHTS

Spotlights have become very popular in the home. Bulbs to fit the units are made in various colours as well as clear. Sizes (wattage) of bulbs range from 75 to 150. The bulbs have internal reflectors. Some have a pressed glass lens and can be used outdoors as well as indoors.

Indoors the lights can be mounted as single units on the wall or ceiling or be fixed to a lighting strip. This is plastic or aluminium extruded trunking, not unlike double curtain rail, containing two circuit conductors which make electrical contact with the special fitting, which is fixed to the light unit instead of a backplate. The lighting units can be positioned anywhere along the track and slid into position. The track can be fixed to the wall or ceiling and has a cable entry hole and terminal block at one end for fixing over a lighting point. Single spotlights have a base plate and have to be fitted to a circular box fixed on the wall or ceiling surface or sunk flush as required. Fixed wiring is not essential.

Spotlights are sold complete with three-core flex and can be supplied from a plug and socket, as can pin-up wall lights. Unless the socket outlet is adjacent to the spotlight it is necessary to fit a longer flex on place of the short flex supplied with the fitting.

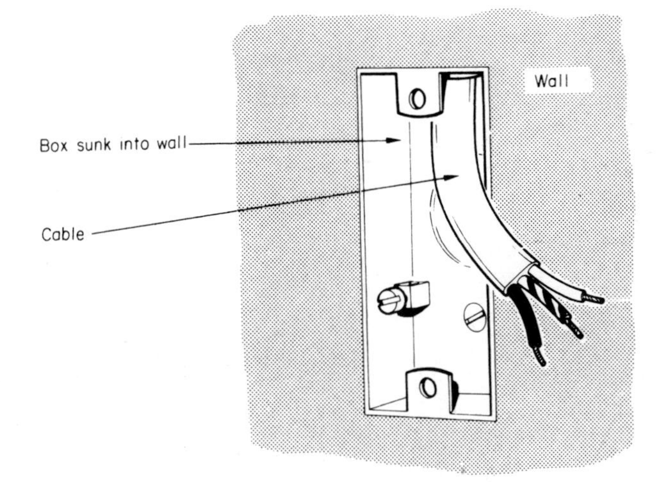

Fig. 43. Wall light fixture.

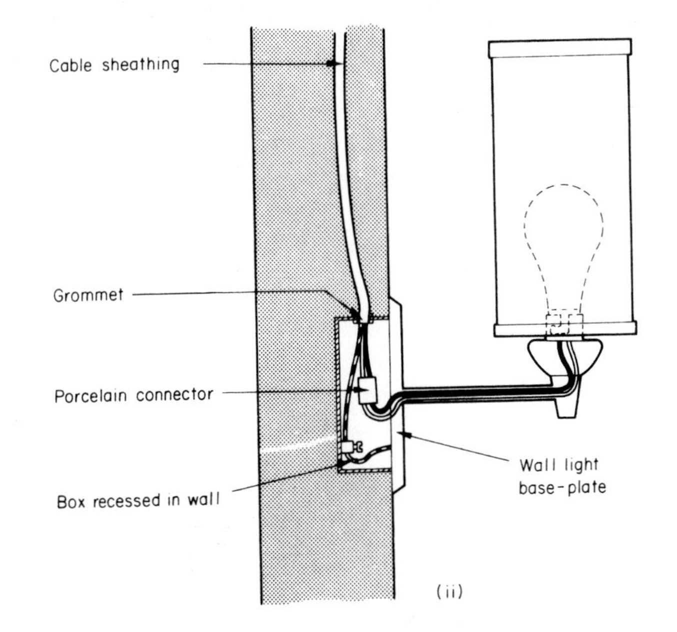

Fig. 44. Installing a wall light.

Fig. 45. Fixed spotlights.

Fig. 46. Spotlight.

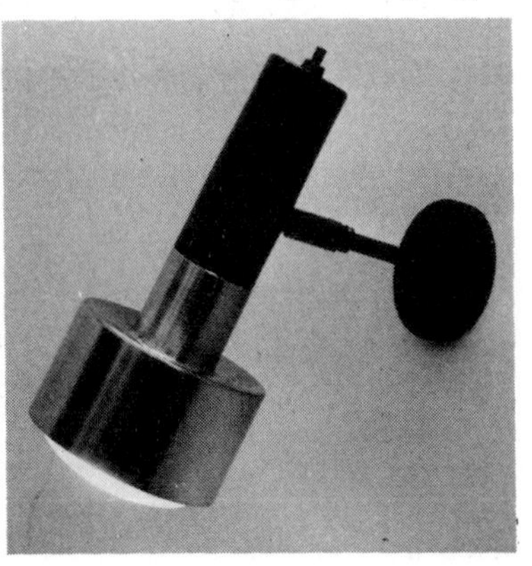

SWITCHES

The majority of lighting switches for fixing to the wall are plate switches consisting of a moulded square plastic plate with the switch behind the plate and operated by a rocker or dolly on the front (Fig. 47). The plate has two screw holes for fixing to the lugs of a box which has 59mm fixing centres. The switch can be mounted on a plastic box fixed to the surface of the wall or to a plaster depth metal box sunk into the wall flush with the surface (Fig. 48).

The traditional height of lighting switches on the wall is 1370mm from the floor to the centre of the switch.

Plate switches having a square plate to fit a one-gang box are made as single switches, two- and three-switch assemblies (Fig. 49). It is therefore possible to have two or three switches situated in the one position on one box, which can be surface or flush-mounted. For two switches a typical position is the hall, one switch controlling the hall light, the other switch

Fig. 47. Lighting switch.

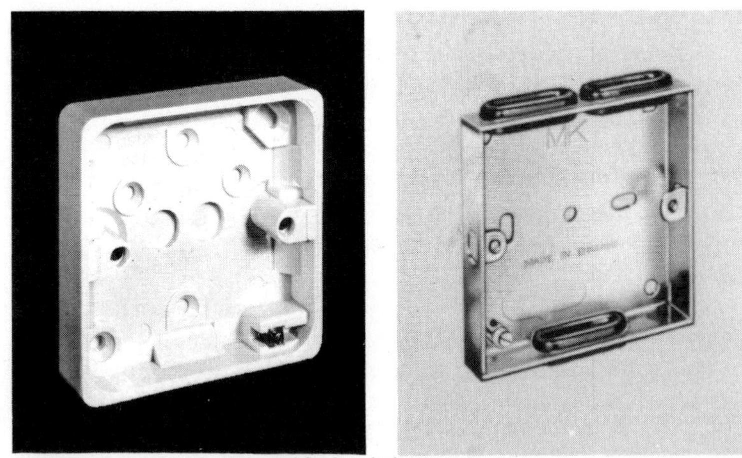

Fig. 48. (a) and (b) Surface and flush switch mounting boxes.

Fig. 49. (a) (b) and (c) Various lighting switches: a) architrave switch; b) two-gang switch; c) three-gang switch.

being the two-way switch of the landing light. A three-gang switch can control a porch light as well.

Where four or more switches are required in the one assembly (rarely in the home), a larger plate is required and the mounting box must be a two-gang.

FIXING SURFACE SWITCHES

Hold the moulded plastic box in position on the wall at a height of 1370mm to the centre of the box (Fig. 50). Mark the box, fixing holes on the wall using a bradawl. Drill or plug holes for a no. 8 wall plug and insert the plugs.

Knock out a section of thin plastic of the box at the top edge to take the sheathed cable or cables. Thread in the cable and fix the box using no. 8 wood screws. Check that the box is level, if necessary using a spirit level. Adjust if required by loosening the screws in the slotted holes and then retightening.

Now strip off the sheathing of the cable, leaving about 12mm within the box. Cut to length. Strip off 12mm of insulation from the ends of the wires and insert in the appropriate terminals, making sure you first slip a length of red PVC sleeving or PVC adhesive red tape over the insulated end of the black conductor.

The red conductor for an on/off switch goes to the terminal marked 'common'. The black conductor with the red identification goes to the remaining terminal. For a two-way switch the connections are as in Fig. 27.

Switches of a two-and three-gang switch assembly are however all two-way switches. Any of these can be used as an on /off switch using only two of the three terminals. The red live wire is connected to the 'common' terminal, but the switch wire is connected to the terminal marked 'one-way'. If not so marked either terminal may be used but if the switch is then upside down (dolly up for ON) the switch wire must be transferred to the other terminal. A single two-way switch may be used for one-way working using the common terminal and one of the others that is usually marked.

Switches and their surface boxes are made in a variety of white or ivory tints and have slight variations in styling. It is therefore necessary to buy boxes of the same make and colour as the switch.

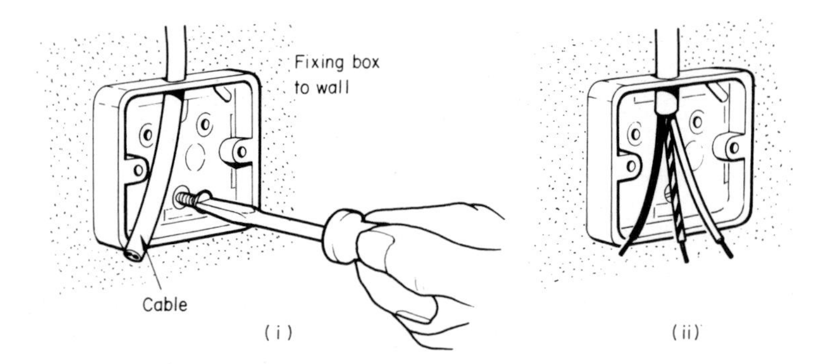

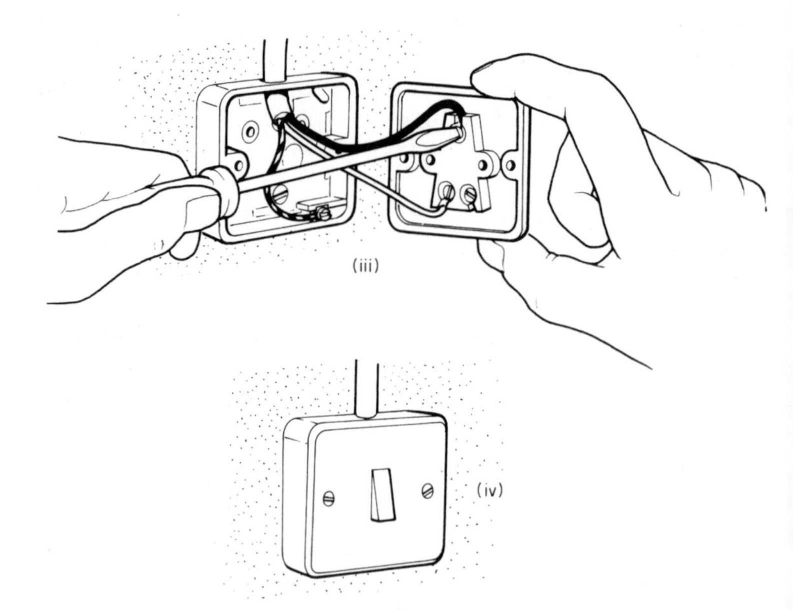

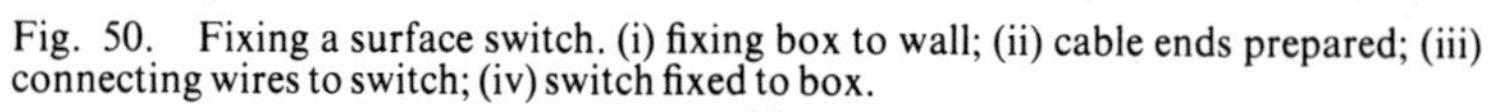
Fig. 50. Fixing a surface switch. (i) fixing box to wall; (ii) cable ends prepared; (iii) connecting wires to switch; (iv) switch fixed to box.

FIXING FLUSH SWITCHES

For these you require a plaster depth box with a PVC grommet in the top edge, which can be reversed if the cable comes up from the floor. If a cable passes through the box a grommeted hole is required at both top and bottom edges.

Hold the box against the wall in the correct position and height. Run a pencil around the external edge of the box to mark the wall for chasing (Fig. 51).

Using a straight edge and a Stanley knife, cut out a square of wallpaper and make an indentation in the plaster to start the chase. Now chop out the plaster, using a sharp chisel which can be an old but sharp wood chisel.

Fit the box in position. If the edges project beyond the wall surface, chop out some brickwork until the box is flush.

Thread the cable into the box via the grommeted hole and fix the box by wood screws in plugged holes. Before tightening the screws check that the box is level and make any necessary adjustments by means of the slotted adjusting holes.

Strip the sheathing and prepare the ends of the wires as for surface-mounted switches and connected the insulated wires to the switch and the green yellow sleeved earth wire to the earthing terminal of the box.

Any make of switch or style and finish of plate switch can be used with a flush box since the switch plate covers it entirely.

DIMMER SWITCHES

A dimmer switch can be used in place of any on/off plate switch and can be fixed to either a flush or surface-mounted box (Fig. 52). It may also be used in place of a two-way switch, but usually in one switch position only, with a conventional two-way switch used at the other switch position. Dimmer switches are also available in two-gang assemblies.

The procedure for fixing a dimmer switch follows that of the conventional switch (Fig. 53).

Dimmer switches are designed for controlling tungsten lighting (ordinary bulbs) only. A version for fluorescent lighting is available but some modification to the circuit is needed. Makers supply details.

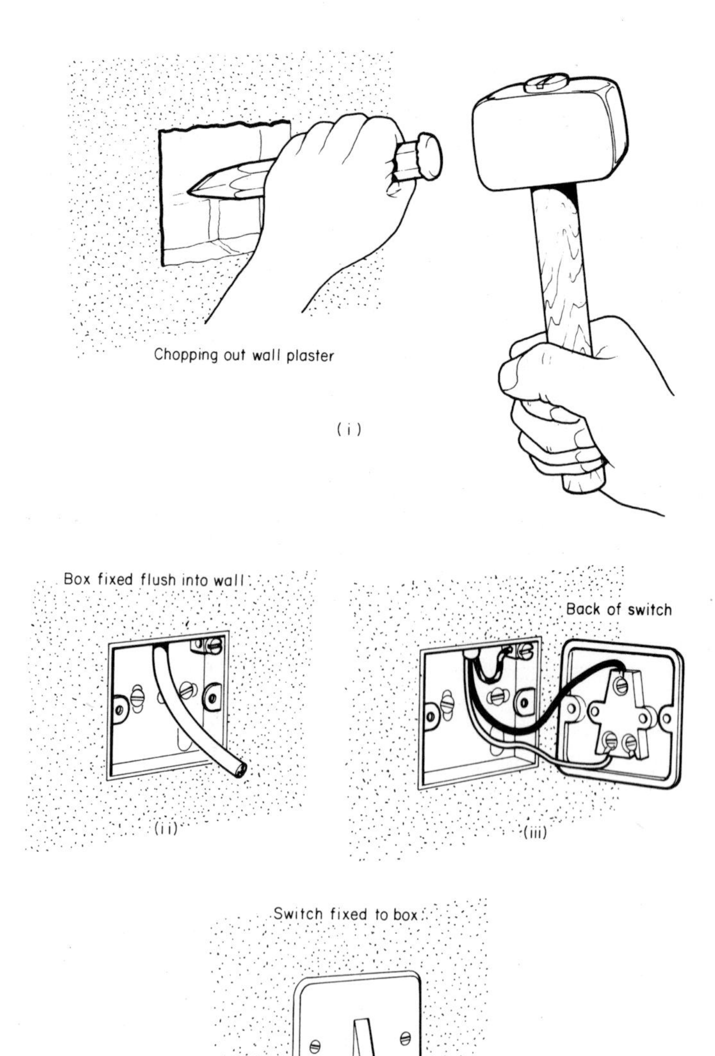

Fig. 51. Fixing a flush switch.

CEILING SWITCHES

The pull-cord ceiling switch (Fig. 54) is obligatory in the bathroom, where a wall switch would be within reach of a person using the bath or shower. (A permitted alternative is to fit a wall switch outside the bathroom door). Ceiling switches are also useful in bedrooms and in many other situations as an alternative to a wall switch.

The ceiling switch is fixed similarly to a ceiling rose and good fixing is essential (Fig. 55). Supports can be either a joist

Fig. 52. (a) and (b) Dimmer switches.

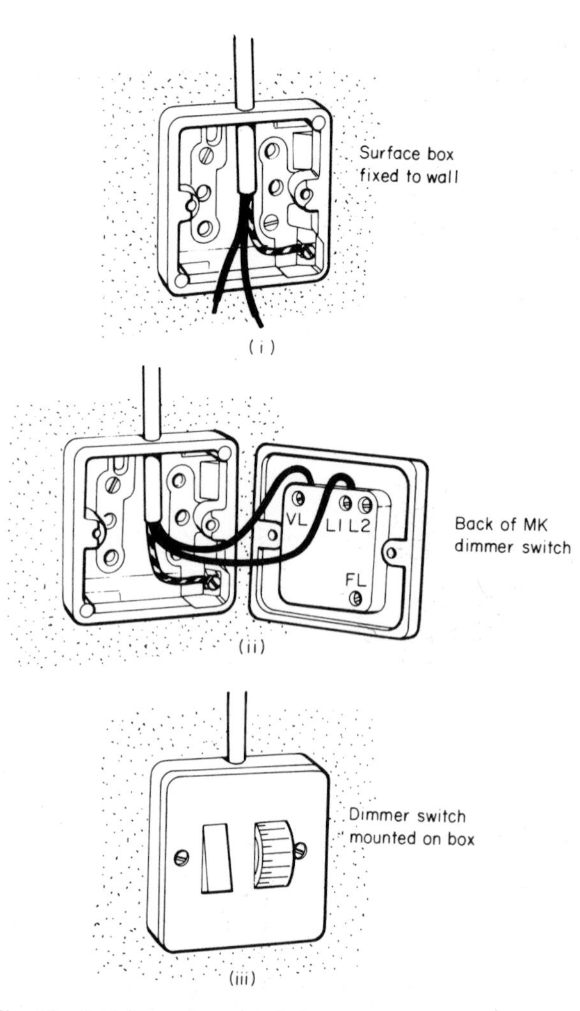

Fig. 53. Installing a dimmer switch.

or a piece of timber secured between two joists as described for ceiling roses. Both one-way and two-way versions are made. Connections are shown in Fig. 55.

Apart from lighting, a cord-operated ceiling switch can be used for controlling a wall heater in the bathroom. This has a current rating of 15A and should be double-pole, an essential requirement when the heater is of the infra-red type having elements exposed to touch. Ceiling switches for heaters are also available fitted with an integral neon indicator, such as is needed for 'black' heating, which has no visual indication that the heater is switched on or off.

New Bathroom Switch.

A 30 amp double pole cord operated ceiling switch with pilot light is used for controlling instantaneous electric shower units which enables the switch to be fixed near the shower instead of outside the bathroom.

Fig. 54. Ceiling switch.

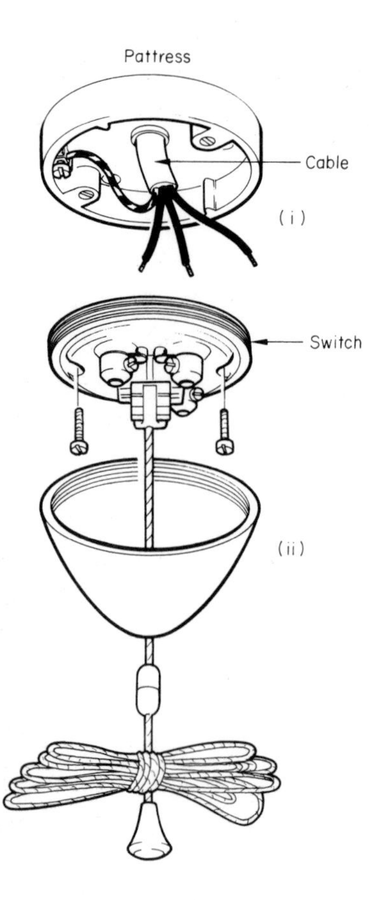

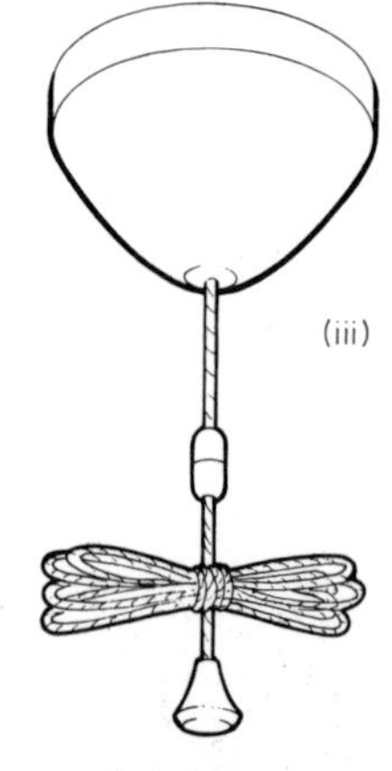

Fig. 55. Installing a ceiling switch.

6

POWER CIRCUITS

RING CIRCUITS

A ring circuit is a multi-outlet circuit that supplies 13A fused plugs and socket outlets and fixed appliances via fused connection units (Fig. 56). The circuit has a current rating of 30A and is protected by a 30A fuse or miniature circuit breaker.

The cable to use for the circuit is 2.5sq.mm twin-core and earth PVC-sheathed cable. This starts at the terminals of a 30A fuse way in the consumer unit, runs through the various rooms and service areas and returns to the same fuse way terminals,

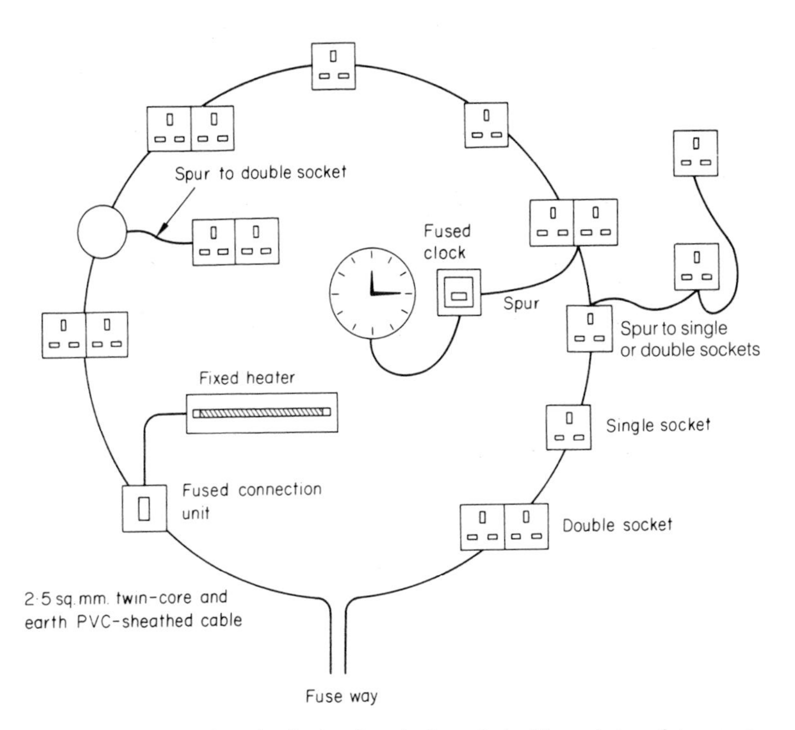

Fig. 56. A typical ring circuit showing single and double socket outlets, spurs branching off the ring cable, and fused connection units supplying fixed appliances (and fixed lighting).

thus completing a ring or loop of cable (Fig. 57). The cable has a nominal current rating of 20A but as it is wired in the form of a ring, which in effect is a double cable, the fuse may have a 30A rating although that is of higher rating than the cable. In addition to the cable forming the ring, spur cables of the same size and type may branch off the ring cable to feed socket outlets and fixed appliances situated in the more remote positions away from the main route of the ring.

A ring circuit and its spurs may supply an unlimited number of socket outlets and fixed appliances but the total floor area served by any one ring circuit must not exceed 100m². Where the floor area of a dwelling exceeds this, additional ring circuits are necessary, one for each 100m² or part thereof.

Since however two ring circuits give double the load capacity of one – 14.4kW instead of 7.2kW – it is preferable to install two ring circuits in a house even where the floor area is less than 100m². Also, should a ring circuit fuse blow, with two ring circuits, only some of the power points will be affected.

I must make it clear that the 7.2kW of load represent the maximum that may be switched on *at any one time* and not the total load of the circuit, which may exceed this amount.

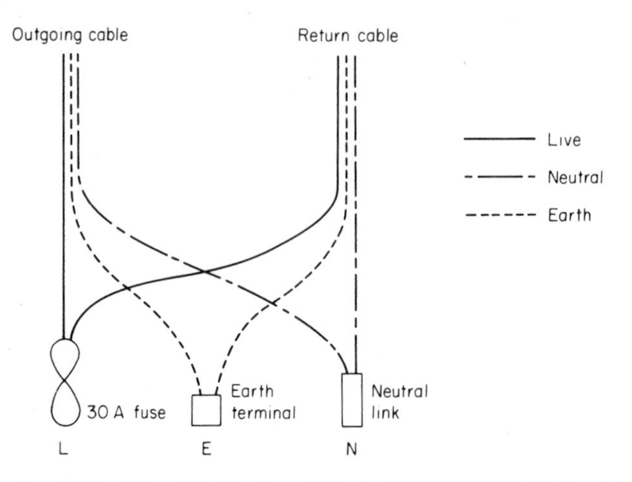

Fig. 57. Connection of ring circuit cables at the fuse way terminals. The two red, the two black insulated and the two earth conductors are joined at the respective terminals.

The number of spur cables that may be connected to a ring circuit must not exceed the total number of outlets connected to the actual ring cable. Any one spur may supply a single 13A socket outlet, or one double or one fixed appliance.

A spur cable may be looped out of a ring socket outlet or be connected to a joint box inserted in the ring cable for the purpose (Fig. 58).

WIRING A RING CIRCUIT

First decide how many socket outlets and fixed appliances the ring circuit is to supply, and mark their positions on the walls. Socket outlets should be fixed at a height of at least 150mm above floor level and at least 150mm above a work surface such as in the kitchen. A preferred height is 300mm, which will bring the socket above the skirting board. In some instances it is desirable to fit sockets at about 900mm above floor level so that you do not have to stoop down to insert and withdraw the plug, or to switch on the socket.

Any outlet on the ring cable may be a double socket, which means you get double the number of sockets without increasing the wiring.

Having decided on the number and positions of outlets you can plan the routes for the cables. On the ground floor of a house and in a bungalow the cable will normally be run under the floor and will rest on the concrete, but where the floor is solid the cable will have to be run behind the skirting, be enclosed in plastic trunking, or be fixed to the surface of the wall. In the upper floors the cables are laid under the floorboards alongside the lighting circuit cables.

THE GROUND FLOOR WIRING

Cables laid on the concrete beneath the floor do not need to be fixed unless they are likely to be disturbed.

It is usually necessary to raise two floorboards in each room, as close to the walls as possible, to assist in fishing out the cables. You will probably also need to raise one the full length of the hall where boards run from the front to the back of the house; otherwise lift three short boards, and also raise one beneath the consumer unit (there should be a loose board here that was used by the electricity and gas people to give access to their pipes and service cable).

At each outlet first chop out some plaster at the top of the skirting board and then, using a long thin cold chisel, remove the plaster behind the skirting to give access to the cables coming up from below the floor. Take care not to damage the skirting boards. The gap behind the skirting board should be wide enough for at least two cables, the outgoing and the incoming, and in some instances for a spur cable also.

Next, insert a length of galvanized wire with a hook on the end to use as a fish wire and a draw wire to draw in the cables. A draw wire is not necessary where a raised board is near a wall, thereby allowing the cables to be pushed up by hand. Where you cannot reach a fish wire from a raised board you will need another length of galvanized wire with a hook on it to fish the first.

It is not necessary to insert fish wires at all points before you start wiring. You can deal with one at a time as you come to it, but you should clear the plaster from behind the skirting boards at every point. Having done this and lifted all the necessary floorboards you can run the cable.

Start at the consumer unit by passing a cable down under the floor and pulling a sufficient length through to reach the first point along the cable route.

Allow sufficient cable to pass up behind the skirting board and to make the connections at the socket outlet, remembering that you will need more cable for sockets at high level than those positioned just above the skirting board.

Return to the consumer unit, allowing enough cable to reach the consumer unit and for the connection within the unit. Cut the cable and take the coil to the first point. Strip off about 150mm of sheathing from the end of this cable and the same amount from the end of the length you ran to the point. Strip off about 100mm of insulation and join the cable to the end of the fish wire. With the aid of a companion pull up the fish wire followed by the cable. Disconnect the fish wire and pull through enough cable for the connections.

Estimate the length of cable to reach the second point and cut the cable from the coil. Follow the procedure you adopted at the first point and carry on until you reach the last point to

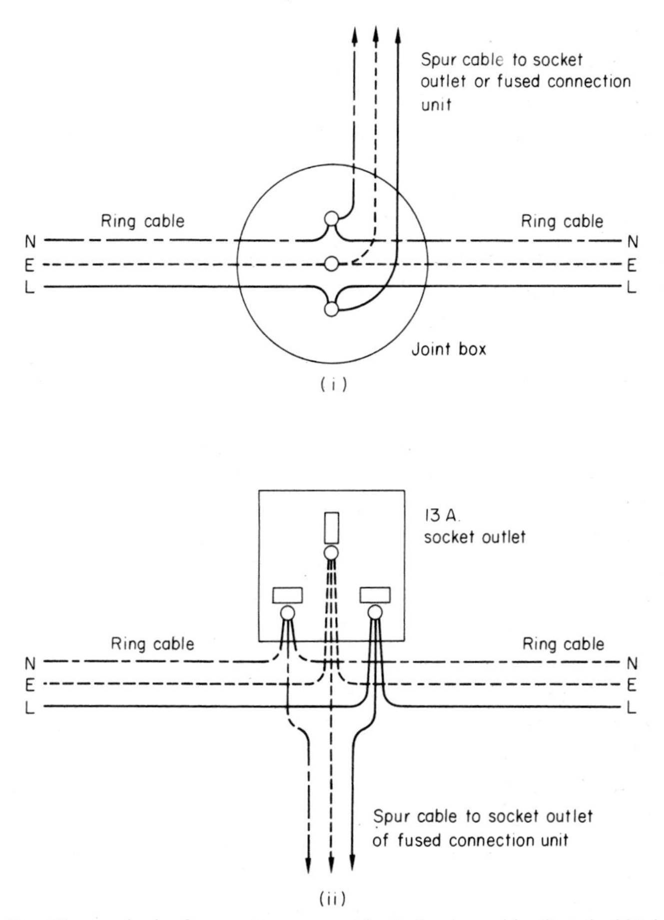

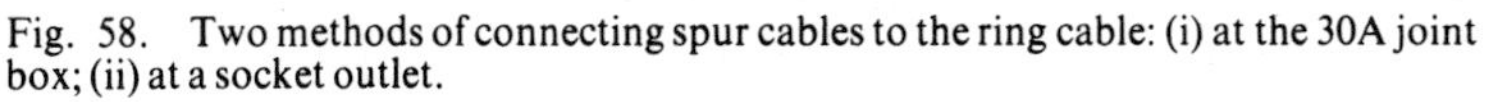
Fig. 58. Two methods of connecting spur cables to the ring cable: (i) at the 30A joint box; (ii) at a socket outlet.

be fed from the ring cable. From this point run the cable back to the consumer unit to complete the ring.

Allow enough cable to pass up to the consumer unit. Push the cable up from the floor alongside the first cable and run both up the wall to the consumer unit using clips to fix them.

If there are any outlets that are to be supplied from spurs, wire these before the floorboards are relaid. A spur will normally be looped out of a socket outlet, so you may need to use the fish wire again.

The remainder of the work will be above the floor so you can replace all boards and carry on with the next floor where relevant.

WIRING THE UPPER FLOORS

In the conventional two-storey house this will be the first floor. Here, floorboards need to be raised as for the ground floor but it will also be necessary to drill joists for cables that have to cross joists.

Decide on the route for the cables and lift the appropriate floorboards. At each point chop out the plaster behind the skirting board as a channel for the cables coming up from under the floor and insert fish wires as you run the cables.

Where a socket outlet is positioned at the end of a board – that is where the wall is at right angles to the boards and not parallel to them – it is necessary to lift the board immediately below the socket outlet position. The joist is usually tight up against the wall so you will have to take the cable through a slot cut in the top of the joist and not through a drilled hole. Then before the floorboard is relaid cut a nick in the end of it to take the cable; otherwise the cable will be jammed up against the wall and will be damaged.

When wiring an upper floor the cable must rise from the consumer unit to the first point and the return cable from the last point must drop back down to the consumer unit. This can be run down the cupboard housing the meter and consumer unit; or, if the unit is not in a cupboard, the cables can be buried in the wall. (For the connections at the consumer unit see 'Installing a Consumer Unit', p. 18).

Having run all the cables and relaid all floorboards, you can now turn to the outlets – the sockets and fused connection units.

Deal first with those at skirting level. The short length of cable from the top of the skirting to the box of the outlet can be sunk in the plaster of the wall using the method described on p. 38. Next, fix the box (Fig. 59) which can be a flush-mounted metal box or a moulded plastic surface box, and connect and fix the accessories as described on page 82.

Where a socket outlet or a fused connection unit is mounted at high level, for example above a worktop in the kitchen, the cable is run up the wall from the skirting. This can be fixed to the surface, enclosed in capping or conduit, or sunk into the plaster using the methods described on page 38.

RADIAL POWER CIRCUITS

A radial power circuit is a multi-outlet circuit of 30A or 20A current rating.

The 30A radial circuit with cartridge fuse may supply a floor area of 50m^2 with 13A socket outlets and fixed electrical appliances installed in different rooms (Fig. 60).

The cable used for this circuit is 4sq. mm twin-core and earth PVC-sheathed cable. The method of wiring is similar to that for the ring circuit in that you start at a 30A fuse way in the consumer unit and feed each point, but the cable terminates at the last point and does not return to the consumer unit. The cable does not however have to be run in a line. Cables may branch out of socket outlets. The maximum number of cables at any outlet should not exceed three or the terminals will not accommodate the conductors.

The 30A radial circuit is especially suitable for the kitchen, either as an alternative to the ring circuit or as an additional circuit to increase the load capacity by 7.2kW. Such a circuit can well supply such auxiliary equipment as extractor fan, cooker hood, waste disposer, central heating circulating pump and sink water heater plus a socket outlet. It is also suitable for a bedsitter or flatlet, where it can supply some lighting as well.

The 20A radial power circuit is wired in 2.5sq. mm two-core and earth PVC-sheathed cable as used for a ring circuit and can supply a floor area of 20m^2 with 13A socket outlets which may be double, or single in different rooms if required. Such a circuit is suitable for a medium-size lounge or bedroom where there is no ring circuit or where an additional power source would be useful should the ring circuit fuse blow.

The 20A radial circuit is useful for supplying a refrigerator and a home freezer, the latter in the garage, where the one circuit is left switched on when the house is left unoccupied for ex-

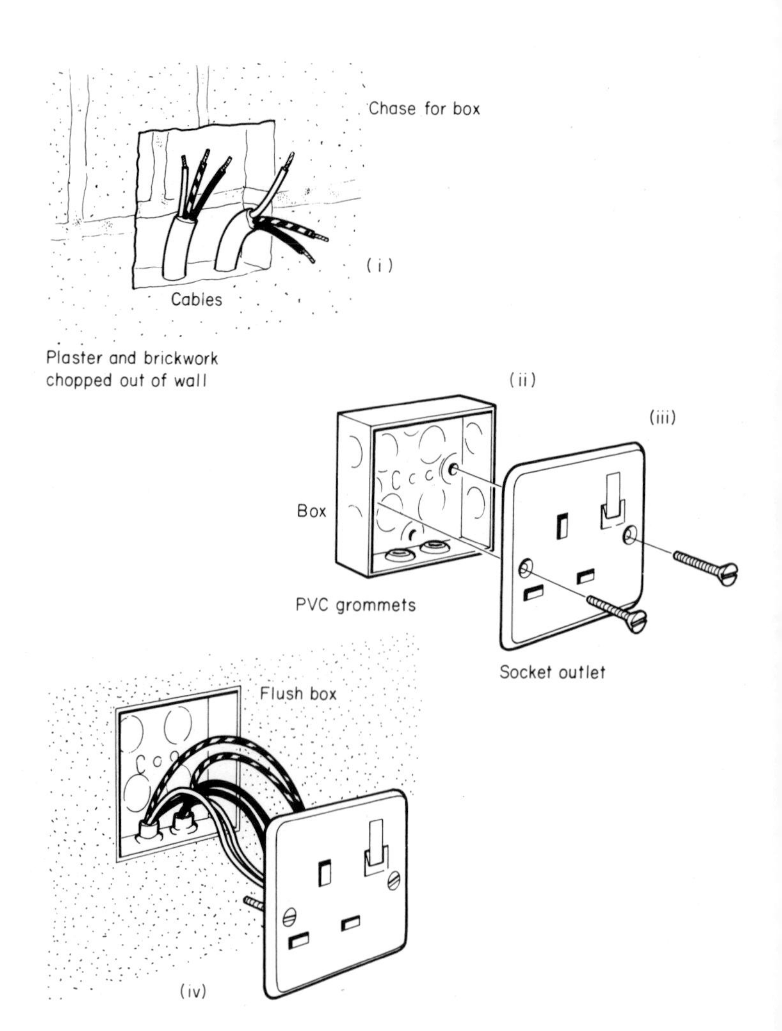

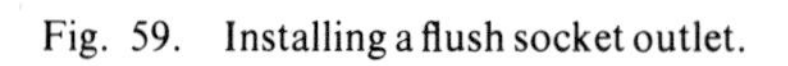
Fig. 59. Installing a flush socket outlet.

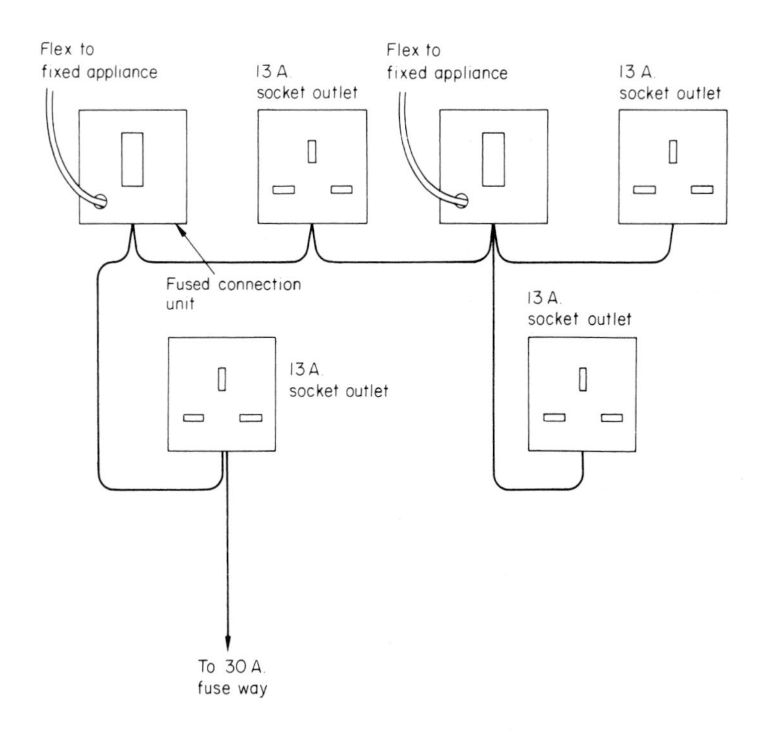

Fig. 60. A typical 30A radial power circuit supplying a floor area of $50m^2$ with 13A outlets consisting of socket outlets and fused connection units.

tended periods and all other circuits are switched off. Switching of individual circuits is made easy by having miniature circuit breakers instead of fuses in the consumer unit.

The circuit is also suitable for security services or for background heating under a time switch control.

7

POWER CIRCUIT ACCESSORIES

The principal accessories used with ring circuits and radial power circuits are: socket outlets; fused connection units (also called fused spur units); fused clock connectors, and flex outlet units. The 30A three-terminal joint box is also an accessory for these circuits.

SOCKET OUTLETS (13A)

There are numerous versions of socket outlet available (Fig. 61): unswitched, switched, and switched with neon indicator, all in single and double socket assemblies.

All the sockets are designed for either surface or flush mounting using the appropriate moulded plastic surface box or flush metal box, the same socket unit being used for either.

FIXING SURFACE SOCKET OUTLETS

The box will be either a one-gang or a two-gang depending on whether the socket is a single or a double. It should be of the same make as the socket so that it matches in both style and colour.

Hold the box in its correct position against the wall and pierce the fixing holes with a bradawl to mark the wall. Drill and plug the wall for no. 8 fixing screws. Knock out the thin plastic in the base of the box for the cable entry. Sink the cables in a chase in the wall and make good the chase with plaster filler. As the plaster needs to set before you fix the socket outlet you can do this job first for all outlets.

Thread the cables into the box and fix the box to the wall in the plugged holes. Cables run to higher mounted sockets can alternatively be fixed to the surface of the wall. Strip the sheathing from the end of each cable leaving about 25mm within the box. Cut the conductors to length (about 150mm) and strip about 12mm of insulation off the end of each. Slip a length of green yellow PVC sleeving over the ends of earth wires. Insert the red conductors into the 'L' terminal of the socket and tighten the grub screws. Insert the black conduc-

Fig. 61. (a) (b) (c) and (d) Various socket outlets: a) single-switched; b) single-switched metal clad; c) double-switched with pilot light; d) triple socket with one 13A and two 2A.

tors into the 'N' terminal and the earth wires into the 'E' terminal (see Figs. 62 and 63).

Now fix the socket outlet to the box using the screws provided. Because the 2.5sq. mm cables have single-strand conductors, these are stiff, and must be carefully placed into the box, but this will not be difficult unless they are shorter than about 125mm.

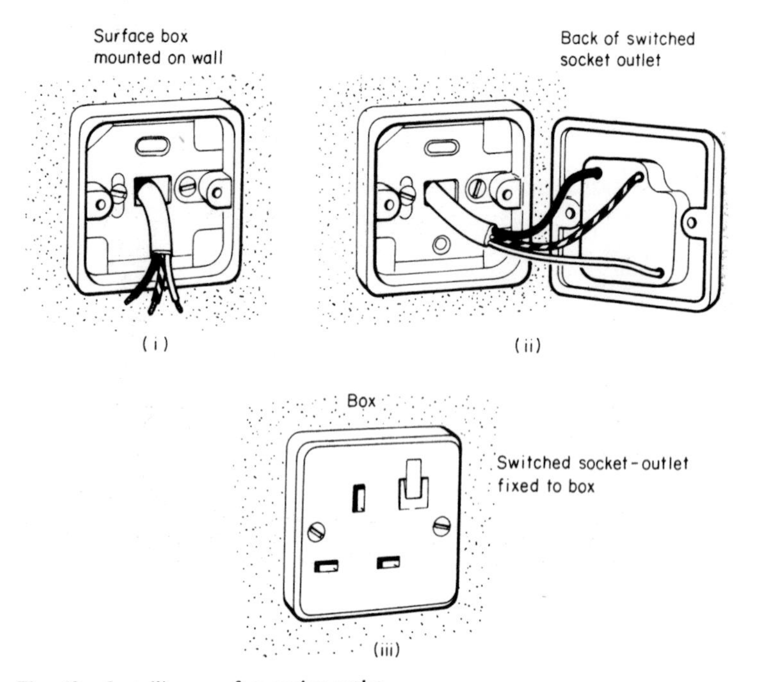

Fig. 62. Installing a surface socket outlet.

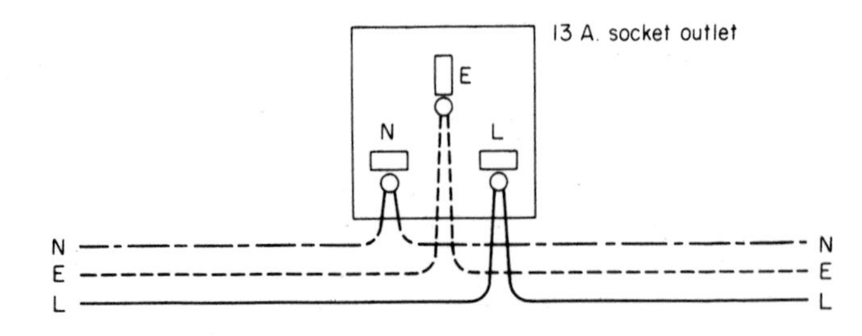

Fig. 63. Connections of a socket outlet to the ring circuit cable.

FIXING FLUSH SOCKET OUTLETS

The procedure is the same as for surface mounting except that the metal box has to be sunk into the wall. To do this, hold the box in position against the wall and run a pencil around the box to mark the wall for the chase.

Chase out the plaster using a sharp cold chisel and place the box in the chase. Measure how much brick is to be cut out for the box to be flush. The overall depth of the box is 48mm. Chop out the brick or breeze making sure the base of the chase is even.

Fit the box in the chase and mark the fixing screw positions. Drill and plug the holes.

Knock out two metal blanks from the lower edge of the box and fit PVC grommets into these knockout holes.

Thread in the cables and fix the box using wood screws.

Prepare the ends of the cables and connect and fix the socket outlet as described for the surface type.

FUSED CONNECTION UNITS

The fused connection unit (or fused spur unit) is intended especially for supplying fixed appliances from ring circuits and radial power circuits.

The basic component is a cartridge fuse of 13A or lower rating and is the same fuse as fitted to fused plugs. The 13A rating is used where the appliance has a loading of from about 700W up to and including 3,000W, the largest appliance that can be used from a ring circuit or multi-outlet radial circuit. Other fuse ratings are usually confined to the 3A, which is used for appliances having loadings up to 720W each, and also for fixed lighting.

The fused connection unit is made in a variety of versions: there are unswitched units used for such purposes as remote lighting points not on a lighting circuit; switched units (these have a double pole switch); and switched or unswitched units with a pilot light (see Fig. 64).

All of these are also made with and without a flexible cord outlet. A flex outlet is required where the appliance is fitted with a flexible cord. It is not required where the appliance is supplied by a fixed cable or supplies fixed lighting.

Fused connection units of all versions are designed for

mounting on both surface and flush boxes as are socket outlets. The same size and type of box is used but they are all one-gang units since all fused connection units are single assemblies. Where two fused connection units are mounted side by side to supply two adjacent fixed appliances, a dual box, not the conventional two-gang box, is used. These boxes are available in surface and flush versions.

Fixing boxes for fused connection units and preparing the cables is done in the same way as described for socket outlets (see Fig. 66). The connections, however, differ depending on the type (fig. 65).

The non-switched version has one neutral terminal, which is used for the circuit cable and also the neutral core of the outgoing flex or cable; two live terminals, one being the circuit cable terminal marked MAINS, the other being the load terminal for the appliance cord or cable; and either one or two earth terminals.

The switched fused spur unit has two N terminals as well as two L terminals. One pair is marked MAINS for the circuit wires; the other pair is sometimes marked LOAD and is for the appliance flex or cable. The earth terminals are the same as for the non-switched version. The reason for the second neutral terminal is because the switch is double-pole and breaks the neutral as well as the live pole.

Fig. 64. (a) and (b) Fused connection units: a) switched with pilot light and cord outlet; b) switched without cord outlet or pilot light.

A flexible cord must be anchored by its sheath within the unit so that it is not pulled out accidentally.

FUSED CLOCK CONNECTORS

This is similar to a fused connection unit with the exception that the fuse section is a two-or three-pin plug to which the flexible cord of the clock is attached (Fig. 67). The plug portion is retained by a screw to prevent its accidental removal which would stop the clock connected to it. The two-pin version is used with all-insulated and double-insulated clocks, the three-pin version for earthed metal clocks.

The wiring connections are the same as for a socket outlet, but as the terminals of some versions are small looping the ring circuit cable at a clock connector should be avoided. Where there are a number of clocks a solution is to supply them from a centrally positioned fused connection unit using 1.0sq. mm lighting circuit cable which will be fuse protected by the 3A fuses in the connection unit.

Clock connectors can be used for supplying individual wall lights.

FLEX OUTLET UNITS

A flex outlet unit is simply a plastic plate containing three ter-

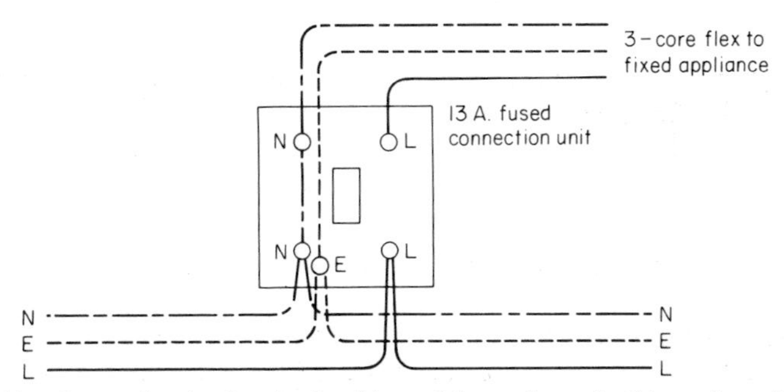

Fig. 65. Connecting the ring circuit cables and the appliance flexible cord to a fused spur unit.

minals, a flex outlet hole and a flex anchorage (Fig. 68). They are used for connecting a fixed appliance to the circuit wiring where no local fuse and no external switch are required, particularly in the bathroom fixed adjacent to a wall fire, towel

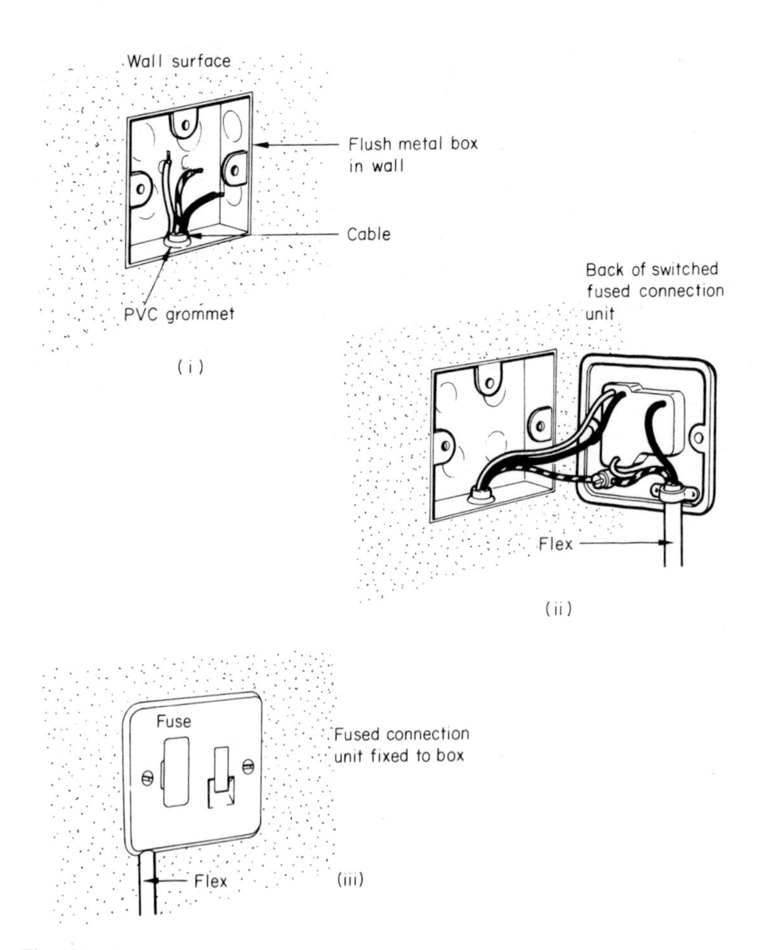

Fig. 66. Installing a fused connection unit.

'rail or other electric heater. The cable feeding the unit is usually a spur from a switched fused connection unit on the ring circuit.

The same type of mounting box is used for the flex outlet unit as for a socket outlet – surface or flush – and the cable connections are also the same except that there are three terminals only, L N and E, the circuit conductors and the flex cores being inserted in the same terminals.

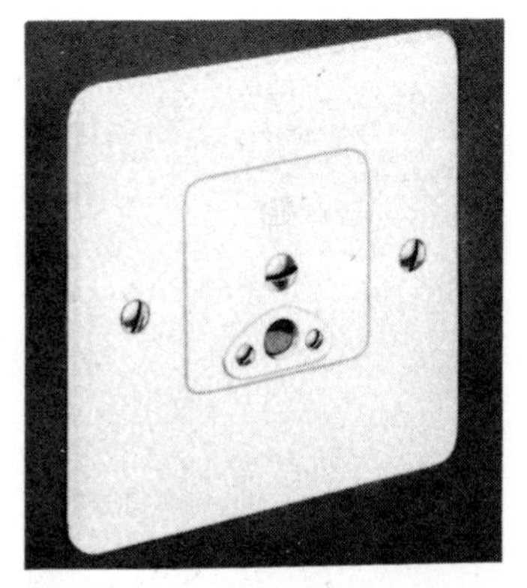

Fig. 67. Clock fused connection unit.

Fig. 68. Cord outlet unit.

SHAVER SUPPLY UNITS

A shaver supply unit is a special socket outlet used only for electric shavers (Fig. 69). It can be connected to a ring circuit, usually (for convenience) to a spur cable.

The shaver unit has an isolating transformer enabling it to be installed in a bathroom provided it is made the British Standard 3052. The unit can be flush-mounted on a flush metal box or surface-mounted on a matching moulded plastic box.

The circuit cable is run into the box and is connected to the terminals of the unit marked L, N and E respectively.

Versions for use in any room *except* the bathroom (Fig. 70) must comply with B.S. 4573. Some of these can be connected to a ring circuit or radial power circuit but some can be connected only to a lighting circuit.

POWER CIRCUIT JOINT BOXES

The 30A, three-terminal moulded plastic round joint box for use with PVC-sheathed cables (Fig. 71) is used mainly for connecting spur cables to a ring circuit but also with radial circuits. The box and its terminals will accomodate four twin-core and earth cables – that is, the incoming and outgoing ring cable plus two spur cables.

Connections of conductors within a joint box are shown in Fig. 71 on p. 90.

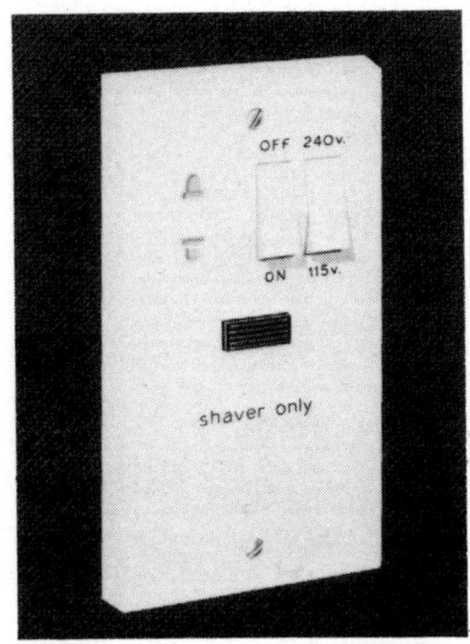

Fig. 69. Shaver supply unit for bathroom.

Fig. 70. Shaver socket other than for bathroom.

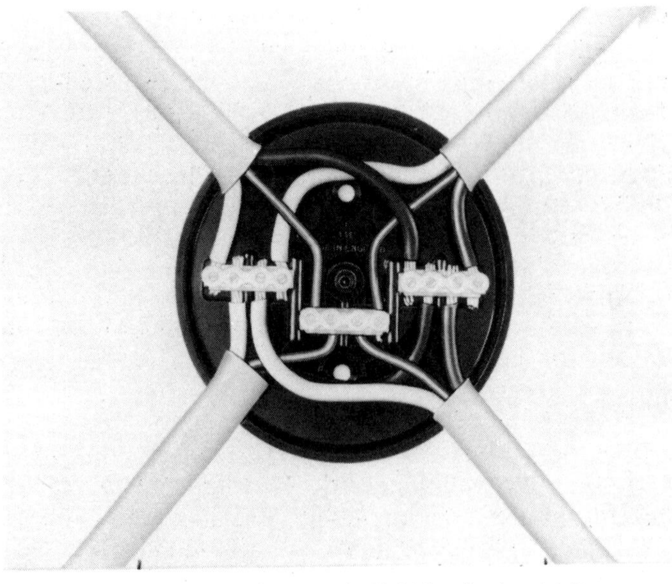

Fig. 71. 30A three-terminal joint box for ring circuit.

8

FLEXIBLE CORDS

A flexible cord is a cable having copper conductors each consisting of a comparatively large number of fine wires to provide the necessary flexibility. The conductors are insulated and are termed cores; a flex has either two or three cores. The sizes (cross-sectional area) of conductors vary according to the current they are required to carry for the various appliances to which they are fitted (see table 3).

The cores of flexible cords are insulated in international standard colours: brown for the live, blue for the neutral, and green-yellow striped for the earth. These have replaced the former colours used in the UK (red, black and green respectively). The old colours can still be used, but it would be an offence for a shopkeeper to sell an electrical appliance or lighting fitting, fitted with three-core flex in the old colours, or in colours formerly used in other countries and fitted to imported appliances. When renewing a flex on an existing appliance, or when extending a flex of the old colours (using approved flex connectors), it is necessary to remember the old colours.

Before replacing a flex on an appliance check that the existing flex is of the correct type and size for the appliance and for the situation where the appliance is to be used. (See table 3 and the section below on 'Flexible cords for appliances'.)

Flexible cord is made in a wide range of types and sizes to suit the appliance or lighting fitting (see Fig. 72). Sizes are given in table 3.

FLEXIBLE CORDS FOR LIGHTING

For plain pendants consisting of a ceiling rose, lampholder and flex, *twisted twin flex* has always been widely used. This is now largely being replaced by PVC sheathed flex, though it cannot be fitted effectively to some of the older lampholders.

Twisted twin flex is available in various types of insulation: clear plastic, through which can be seen the copper or tinned

Table 3 Flexible cords

Size (sq.mm)	Current rating (amps)	Application
0.5	3	Lighting fittings
0.75	6	Lighting fittings, small appliances
1.0	10	Appliances up to 2,000W
1.25	13	Appliances up to 3,000W
1.5	15	Appliances up to 3,600W
2.5	20	Appliances (4,800W maximum)
4.0	25	Appliances (6,000W maximum)

copper conductors, rubber insulated, red and black cores, covered with cotton or silk braiding in various colours and opaque plastic insulation, usually white. Either of two sizes is used for pendants, 0.5sq. mm (3A) and 0.75sq. mm (6A).

Another twin flex, termed *parallel twin flex*, is used for wiring lighting fittings and also for some small appliances such as electric clocks and shavers. This flex is made in the same two sizes as twisted twin flex.

PVC circular sheathed flex is used for plain pendants with the modern loop-in ceiling rose and lampholder. The sheath is usually white, but black sheath is used in some pendants including modern multilight pendant clusters.

This flex is made in two- and three-core versions. The two-core is used with plastic lampholders; the three-core with metal lampholders and fittings. The core colours of three-core flex are brown (live), blue (neutral) and green-yellow striped (earth). The cores of two-core flex are brown and blue respectively, the green-yellow earth core being omitted.

Where the bulb in a pendant is 100W or larger and heat at the lampholder is likely to become excessive, *heat-resisting flex* should be used; the lampholder also should be heat-resistant – either heat-resisting plastic or metal with a ceramic interior. A metal lampholder for a plain pendant should have an earth terminal and be wired with three-core flex.

FLEXIBLE CORDS FOR APPLIANCES

It is important that the correct type of flex, as regards size and

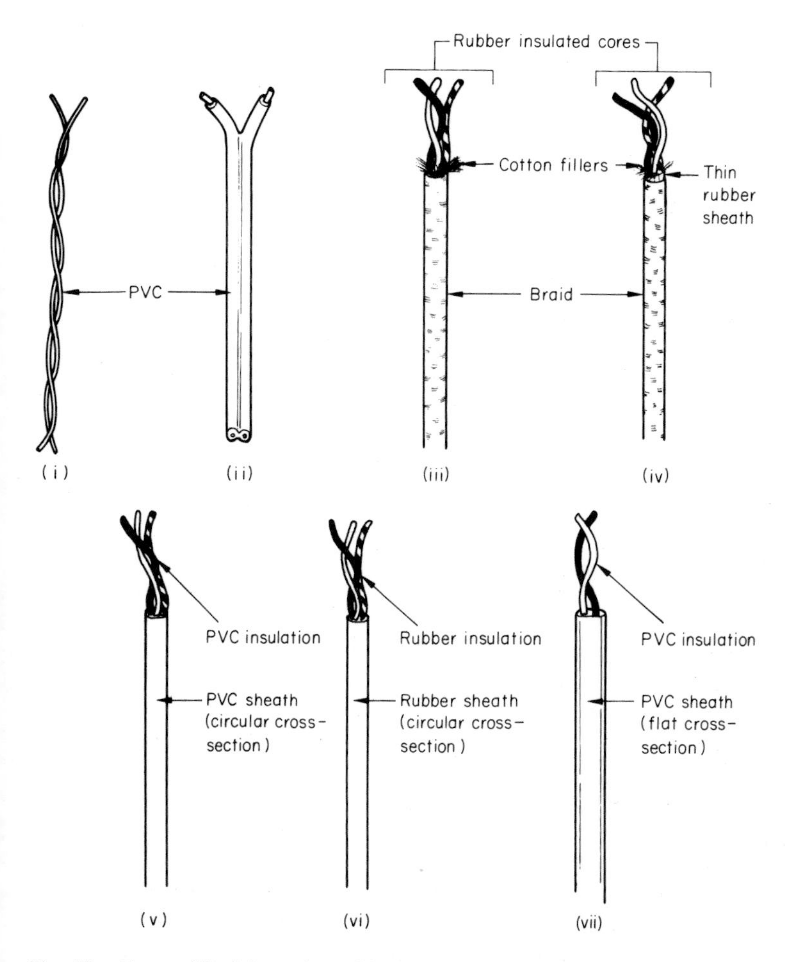

Fig. 72. Types of flexible cords used in the house: (i) twisted twin flex; (ii) parallel twin flex; (iii) circular braided flex; (iv) unkinkable flex; (v) circular PVC-sheathed flex; (vi) circular rubber-sheathed flex; (vii) flat twin-core PVC-sheathed flex.

current rating, be fitted to an appliance.

Appliance flexes are of the following types (see Fig. 72):

(i) circular braided flex;
(ii) unkinkable circular sheathed flex;
(iii) circular PVC-sheathed flex;
(iv) circular rubber-sheathed flex;

(v) flat-twin sheathed flex;
(vi) parallel twin flex;
(vii) heat-resisting flex.

CIRCULAR BRAIDED FLEX

This flex has vulcanized rubber-insulated cores plus cotton fillers to provide the circular cross section. The cores (in standard colour coding) and fillers are enclosed in a two-colour cotton braiding.

This flex is fitted to a wide range of domestic appliances apart from electric kettles and irons (many makes of kettles and irons are sold already fitted with this flex, however).

UNKINKABLE CIRCULAR SHEATHED FLEX

This flex has rubber-insulated cores and cotton fillers enclosed in a light covering of vulcanized rubber and an overall two-colour braid. The type is confined to three-core flex. It is fitted to electric kettles, electric irons, coffee percolators and similar appliances.

CIRCULAR PVC-SHEATHED FLEX

This flex has PVC-insulated cores – two- and three-core – enclosed in PVC sheathing moulded in circular cross-sectional area, therefore requiring no fillers.

This is a very robust flexible cord for indoor and outdoor use. Appliances fitted with it include vacuum cleaners, floor polishers, drills and other power tools, mowers, hedge-trimmers and other garden tools. The flex will withstand much flexing without kinking and will also stand up to fairly rough treatment. In addition it is fitted to an increasing range of indoor appliances, including electric heaters, in place of circular braided flex and is the circular sheathed flex fitted to lighting pendants described earlier.

Sheathing colours are grey, black, white, orange and safety yellow. White and (sometimes) black are fitted to lighting pendants and other lighting fittings; black and grey to home appliances, and orange and yellow to garden tools so that the flex can easily be seen in long grass and hedges to reduce the risk of cutting it.

CIRCULAR VULCANIZED RUBBER SHEATHED FLEX

This flex has rubber-insulated cores and together with fillers is enclosed in black or grey vulcanized rubber sheathing. It is fitted to a wide range of electrical appliances as an alternative to PVC circular sheathed flex, especially in situations subjected to fairly wide temperature variations.

FLAT TWIN PVC-SHEATHED FLEX

This flex has PVC-insulated cores enclosed in PVC sheathing of flat cross section. It is used for double-insulated appliances as an alternative to circular sheathed PVC flex.

PARALLEL TWIN FLEXIBLE CORD

This is twin flex having PVC-insulated cores but no sheath and usually of figure 8 cross section. Both cores are of the same colour, usually opaque white, but for identification one core is sometimes ribbed, this being essential for appliances fitted with single-pole switches such as thermostats or other single-pole devices that must be connected to the live pole.

It is fitted to small-current all-insulated or double-insulated appliances such as electric clocks and also, as already mentioned, to some lighting fittings.

HEAT-RESISTING FLEX

This is a three-core flex having butyl – or EP-rubber compound heat-resisting insulation. It is fitted to immersion heaters, water heaters and space heaters in the vicinity of excessive temperatures which would adversely affect PVC or vulcanized rubber. It is also fitted to lighting fittings subjected to high temperature.

REPLACING A FLEX

Locate the terminal block on the appliance and remove the cover. This and the type of terminal box varies considerably. In some cases you may have to partly dismantle the appliance; in others merely to release a couple of screws (Fig. 73).

Remove the old flex carefully, noting to which terminal each core is connected. If the terminals are not colour-coded do this with blobs of paint the same colours as the old flex cores.

Pass the new flex through the grommet of the cord entry, secure the sheath to the cord grip or other achorage and connect the cores to the terminals.

The brown core must be connected to the live terminal in place of the former red core; the blue core of the new flex to the neutral terminal in place of the former black core; and the green-yellow core to the earth terminal.

FITTING A PLUG TO THE FLEX

The 13A fused plug has identical pins and pin spacings for all makes, but the terminal and cord anchorage arrangement differ somewhat.

Open up the plug by releasing the cover screw (Fig. 74). Remove the fuse and release the screws of the cord anchorage, except in the new MK plug which has no anchor screws but a special cord grip.

Strip 75mm of sheath from the end of the flex and place the end of the sheath over the cord grip position of the plug.

Place the cores at the respective terminals and trim them for length. Strip about 10mm of insulation from the end of each and insert the bared ends into the terminals in the following preferred sequence.

Insert the green-yellow core into the earth terminal first so that you won't mistakenly insert the live into this terminal. The earth terminal is the top centre pin, which is larger than the other two.

Next insert the blue core in the terminal on the left and marked 'N'. Insert the brown core in the live terminal, which is on the right next to or under the fuse and marked 'L'.

Replace the cord grip and tighten the screws, making sure it is the sheath and not the unsheathed cores that is under the grip.

Arrange the cores in the grooves of the plug and refit the cover. With the MK plug, the sheath of the flex is merely pressed into the cord grip.

PLUG ADAPTORS

A multiplug adaptor enables two or more appliances and portable lamps to be run off single socket outlets (Fig. 75). Some patterns allow plugs of other sizes and types to be operated from the standard 13A socket. Where these are of lower

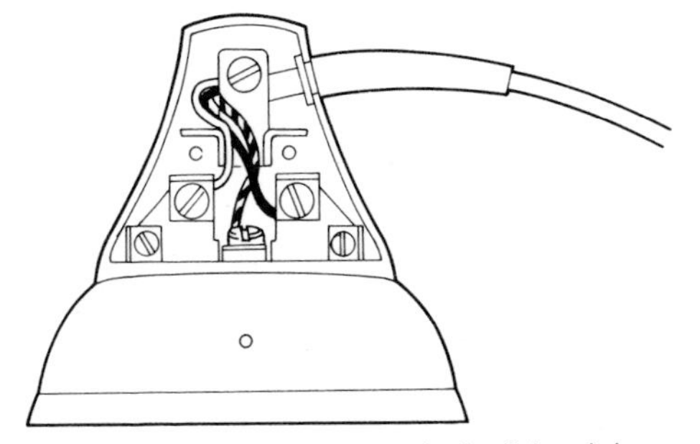

Fig. 73. Flex connected to terminals of electric iron.

current rating the adaptor is fitted with a fuse.

Although convenient, adaptors tend to result in numerous long flexible cords extending from socket outlets. There is also a risk of a socket being overloaded. It is better to install more socket outlets than to rely on adaptors.

FLEXIBLE CORD EXTENSIONS

These are used to provide a longer flex to a power tool, a mower, a hedge-cutter or similar appliances; and also for handlamps and inspection lamps (Fig. 76).

A flex extension consists of a length of circular PVC-sheathed flex with a plug on one end and a trailing socket on the other. With some the flexible cord is wound on a reel and the socket outlet is fixed to the reel (Fig. 77).

Extensions are available with three-core and with two-core flex. The three-core extensions may be used with double-insulated appliances but two-core extensions may be used *only* with double-insulated appliances because earthed appliances and tools, with their exposed metalwork, require three-core flex for an extension to provide continuity of earthing. When you buy a flex extension to use with both earthed and double-insulated appliances it is advisable to buy one fitted with three-core flex so that there is no risk of using a two-core extension with the earthed appliance or tool.

For garden appliances choose an extension having orange or yellow sheath. If you use an extension attached to a reel

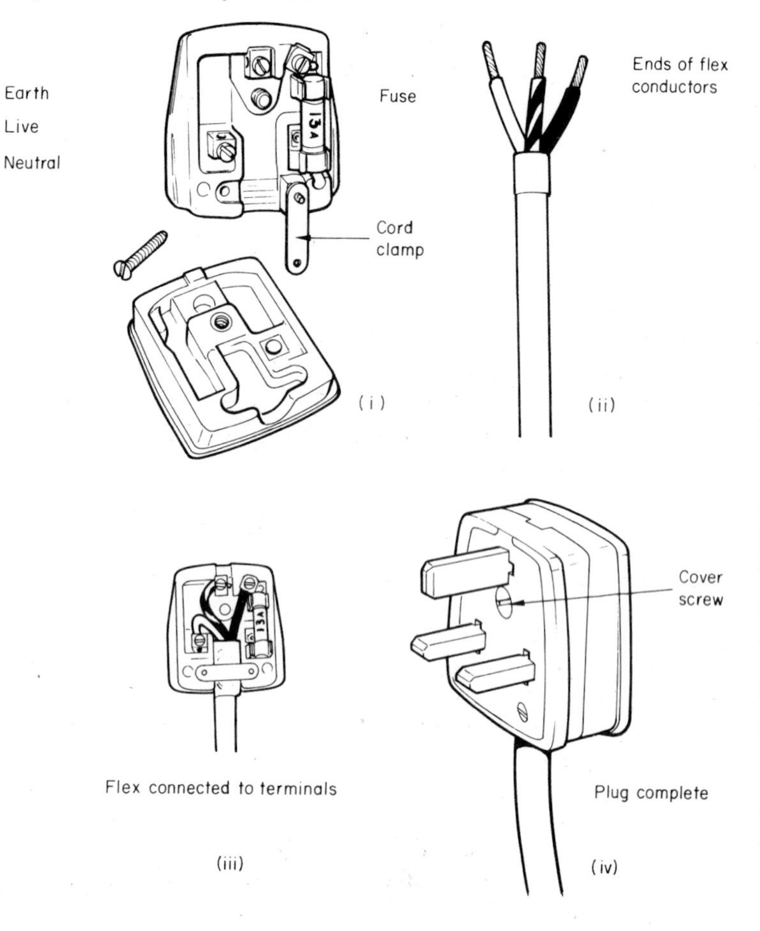

Fig. 74. Connecting flexible cord to a fused plug.

Fig. 75. 13A plug adaptor.

Fig. 76. Single-trailer socket; and multi-outlet socket distribution unit.

Fig. 77. Flex extension on reel.

always run all the flex off the reel; otherwise the flex may overheat and start a fire. This is because the current rating of the flex is based on the flex being used in 'free air'.

9

INSTALLING A COOKER

An electric cooker can be free-standing, as are the majority, or a split-level type built into the kitchen equipment. For either type one circuit only is required: a twin-core and earth PVC-sheathed cable originating at a fuse way in the consumer unit and terminating at a special cooker control unit. This unit must be fitted within 1830mm of the cooker it controls (Figs. 78).

The current rating of a cooker circuit is usually 30A. This covers a four-hotplate family-size model having a total loading of 12kW. For larger models, or where most of the hotplates and oven will normally be in use at the one time, a 45A circuit is installed. This is wired in a larger cable and originates at a 45A fuse way in the consumer unit. As stated earlier, only some models of consumer unit have facilities for a 45A fuse unit, so if this will be required it is necessary to install one that has. An alternative is to fit a separate 45A main switch and fuse unit especially for the cooker circuit and have it connected to the meter by the electricity board.

The cooker control unit is normally fixed at a height of 1500mm and adjacent to or above the cooker (Fig. 79 and 80). The traditional unit contains a socket outlet for a kettle, but as this encourages the housewife to stand the electric kettle on the cooker hob where the flex is likely to trail over a switched-on hotplate, a socketless cooker control switch has now been introduced. This is recommended where the control switch is located above the cooker or sufficiently close to it to enable the kettle to be used on the hob.

For a new circuit the control switch, if it is to contain a kettle socket outlet, should be fitted to one side of the cooker but within the regulation 1830mm distance. The kettle can then be used on the kitchen worktop or a table. Where the circuit is already installed and the cable is buried in the wall and terminates over the cooker, a socketless cooker control switch is a solution. This is a double-pole switch of 45–60A rating and available with and without pilot light.

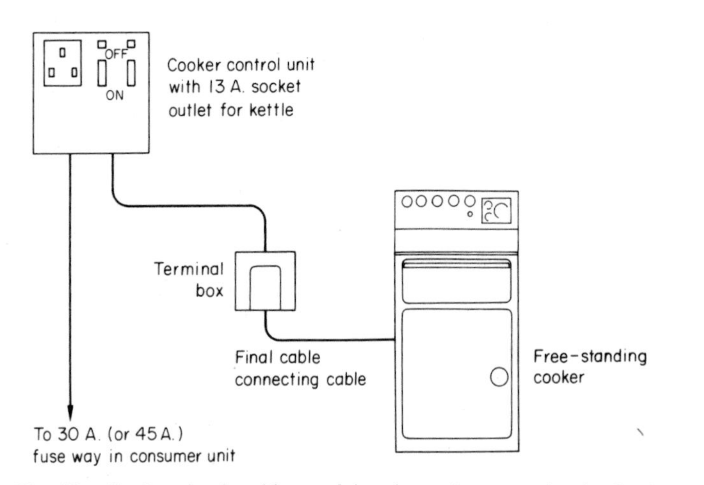

Fig. 78. Cooker circuit cable supplying the cooker control unit of a freestanding cooker and the cable to the cooker with an intervening terminal box.

Fig. 79. Cooker control unit with 13A socket and pilot lights.

Fig. 80. Cooker control switch with pilot light.

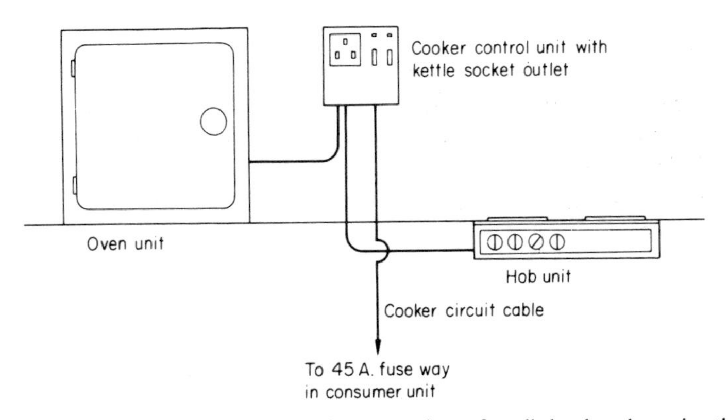

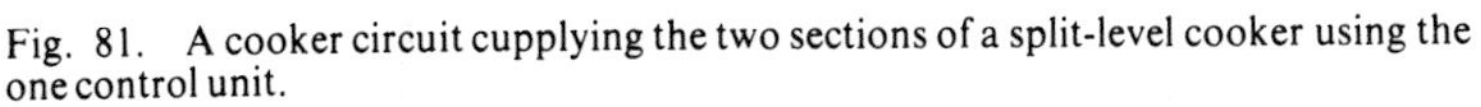
Fig. 81. A cooker circuit cupplying the two sections of a split-level cooker using the one control unit.

CONNECTING THE COOKER TO THE CONTROL

A length of the same size (4 or 6sq. mm) and type of cable as that used for the circuit is used to connect the control unit to the cooker. As it is necessary to move a free-standing cooker out from the wall to clean behind, the lower section of this cable cannot be fixed, but the upper part can be secured to the wall. A terminal box is fitted behind the cooker at the cooker cable outlet level. From the control unit to the terminal box a length of cable is run down the wall; from the terminal box a trailing cable runs to the cooker. The upper section can be fixed to the wall, buried in the plaster, or enclosed in buried conduit.

An alternative is to use a cable-through box having no terminals. The advantage of the terminal box is that the cooker can be removed simply by disconnecting the short length of attached cable from the terminal box.

SPLIT-LEVEL COOKER CIRCUITS

A split-level cooker comprises two sections, a hob unit and an oven unit, often fitted some distance apart. One control unit can supply both sections provided each will be no more than 1830mm from the control. By fitting the control unit half-way between the two sections, therefore, they can be up to 3.7m apart, which gives scope in designing the kitchen.

The control unit is supplied by the one circuit cable and each cooker section is supplied from a separate cable running from the control (see Fig. 81). These cables must be of the same size and can be of the same type of cable as used for the circuit. As the two cooker sections are fixed, the cables can be sunk in the wall throughout their length.

The cooker control may have a kettle socket outlet if required.

10

INSTALLING A WATER HEATER

There are three principal types of electric water heaters from the circuit aspect: (i) storage; (ii) instantaneous; (iii) immersion heaters.

STORAGE WATER HEATERS

These range in capacity from 6.5 litre, which is fitted over the sink, wash basin and other single outlet positions, to 100 and 150 litre, which supply all the domestic hot water requirements in the house. The electric loadings of these water heaters range from 750W to 3,000W and can therefore be supplied from a 15A circuit.

The small sink storage water heater (Fig. 82) has a loading of from 750W to 3,000W depending on the model. These can be supplied from a separate circuit or from a spur of the ring circuit and can be connected to a 13A fused plug and socket outlet or from a 13A fused connection unit. When fitted over a wash basin the same type of outlet can be used except that in the bathroom it must not be a plug and socket and the switch must be out of reach of the bath. An alternative to a fused connection unit is a cord-operated ceiling switch and a flex outlet unit fitted next to the water heater.

The larger size of water heater, although it has a maximum loading of only 3kW, should not be supplied from a ring circuit spur but from a separate circuit. This is because, being a fairly continuous load, it would deprive the ring circuit of much of its load capacity required for portable appliances. The separate circuit can have a current rating of 20 or 15A and its control switch should be a double-pole 20A switch with cord outlet and pilot light, and a plate engraved 'Water heater'.

The circuit cable for a storage water heater whether a spur or a separate circuit, is 2.5sq. mm twin-core and earth PVC-sheathed cable.

INSTANTANEOUS WATER HEATERS

Instantaneous water heaters heat the water as it passes through

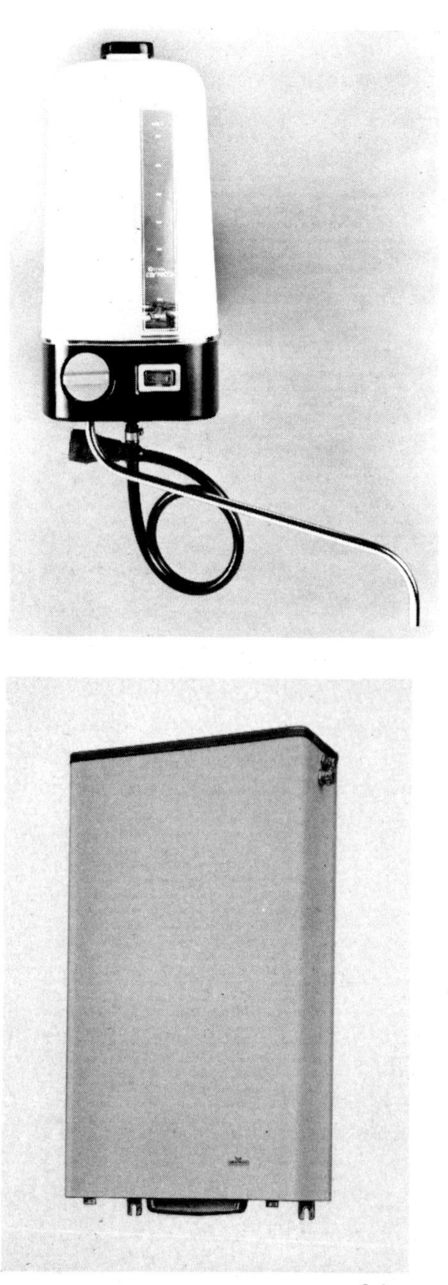

Fig. 82. Creda Corvette sink water heater.

Fig. 83. Multipoint storage water heater with integral cold water ball valve cistern.

the unit. Generally they have no storage capacity though some models have a small residual amount surrounding the element. Because the water is heated as it flows from the mains, the electric loadings of the units have to be comparatively high in order to provide a useful outflow at usable temperature.

The loadings range from 3kW to 7kW. A 3kW size is confined to hand washing at the wash basin; a 4kW or 5kW is fitted over the kitchen sink, and a 5kW or a 6kW model supplies a shower (Fig. 84). The 7kW model can supply two outlets by means of a special two-way faucet valve.

The 3kW model requires a 15A circuit which can be a spur from a ring circuit. Except in the bathroom, the outlet can be a plug and socket. In the bathroom the outlet can be either a switched fused connection unit or a ceiling switch and separate flex outlet combination as for a small storage water heater referred to earlier.

The 4kW or 5kW instaneous water heater is supplied from a separate 20A circuit wired in 2.5sq. mm cable from a 20A fuse way in the consumer unit. The 6kW and 7kW model each require a 30A circuit wired in 4sq. mm cable from a 30A fuse way in the consumer unit.

The control switch of these water heaters must be out of reach of a person using the shower and, if installed in a bathroom, out of reach of a person using the fixed bath. In some instances this means that the switch must be fitted outside the bathroom or fitting a cord operated switch.

I should point out that an instantaneous water heater is not suitable for supplying hot water to a fixed conventional bath. The rate of outflow of water at the required bath temperature of 110° F is too low for even the 7kW model to fill a bath. In fact, a loading of about 18kW would be needed to fill a bath in an acceptable length of time.

Shower units can be bought complete with a shower cubicle which can be fitted in any room in the house including the kitchen, the landing or a hall cupboard.

IMMERSION HEATERS

An immersion heater is installed in the conventional hot water tank or copper cylinder either as the sole provider of hot water to all hot water taps or to operate in conjunction with a back-

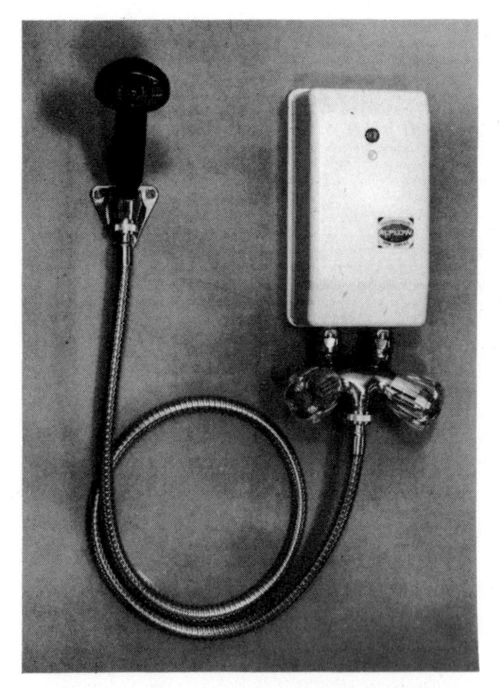

Fig. 84. Instantaneous water heater for shower.

fired boiler or independent boiler. It is available in single-element and twin-element versions but its maximum load demand does not exceed 3kW. With a twin-element type, where each element has a loading of 3kW, a changeover switch in the circuit or the heater head prevents both elements being switched on together (Fig. 85).

The circuit for an immersion heater (Fig. 86) has a current rating of 20A wired in 2.5sq. mm cable from a 20A fuse way in the consumer unit. The circuit cable terminates at a 20A double-pole switch which has a cord outlet and can have a pilot light; if desired the switch can be engraved 'Water heater'. The flexible cord connecting the immersion heater to the control switch should be heat-resisting, using a material such as butyl-rubber or EP-rubber compound for its insulation and sheathing.

For a twin-element immersion heater, a combined double-pole switch and change-over switch is used, though some models have a change-over switch on the heater or a remote

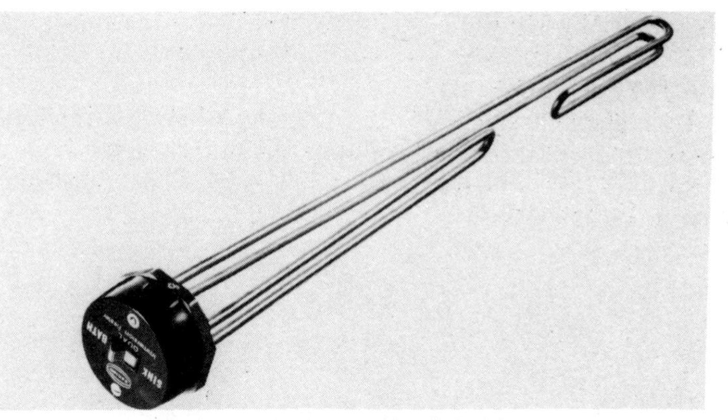
Fig. 85. Dual immersion heater with switched thermostat unit.

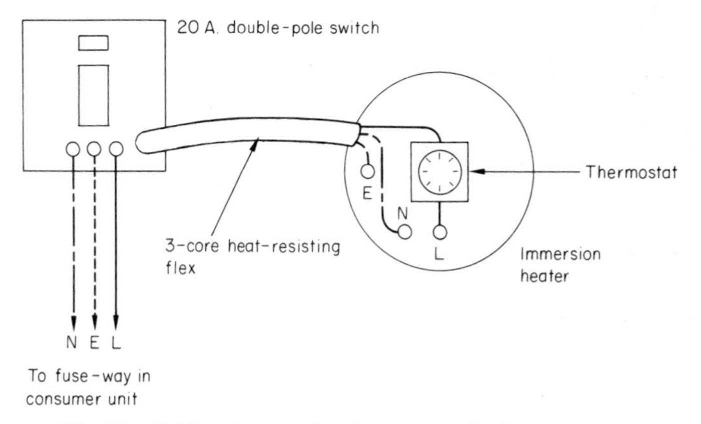

Fig. 86. Cable and connections for an immersion heater.

switch fitted in the kitchen. The two-element unit has a long element for heating the whole of the contents for a bath and a short element to keep a few gallons hot for normal hot water requirements.

An immersion heater can be controlled from two positions: at the heater and, say, in the kitchen. The switch at the heater is a master switch and the one in the kitchen an auxiliary switch, but the immersion heater can be switched on and off from either position and both switches have a pilot light.

Special controls, and sometimes a second immersion heater,

are required when the heater is to operate on the cheap night rate of the 'Economy 7' tariff and additional hot water is required during the daytime.

Arrangements must always be made, where the tank cupboard opens into the bathroom or the tank is actually in the bathroom, that the control switch cannot be reached by anyone using the bath.

11

INSTALLING STORAGE HEATERS

Much of the electric heating in the home is by means of portable electric heaters plugged into socket outlets of the ring circuit as and when required. Other electric heating consists of fixed heaters connected to the ring circuit via fused connection units. No special wiring is needed except where heaters are to be controlled by room thermostats and/or time switches, for example skirting and other perimeter heaters used for background heating. These can be supplied from separate radial circuits or from the ring circuit via fused connection units.

Electric storage heating differs from all other forms of electric heating in that the circuits are time-controlled, so that the heaters take electric current only during a seven-hour overnight period when electricity is supplied at a reduced rate under the 'Economy 7' or special off-peak tariff.

Storage heaters are therefore supplied from a second consumer unit, connected separately to the mains electricity supply but to the same meter as the other electricity services. This meter is a dual-rate meter and there is a time switch to effect the change of rate and usually to switch the storage heating on and off night and morning.

As all storage heaters take current at the one time with no diversity of use between them, as there is with direct-acting heaters, each heater must be supplied from a separate circuit (see Fig. 87). If for example you install three storage heaters, you will require a three-way consumer unit, but if an immersion heater or a 20 or 30gal storage water heater is also to operate on the time-controlled supply an additional fuse way is needed. This would mean a four-way consumer unit for three storage heaters and the water heater.

Current ratings of the circuits are 20A, each circuit being wired in 2.5sq. mm twin-core and earth PVC-sheathed cable from a 20A fuse way in the time-controlled consumer unit. The circuit outlet at each storage heater should be a 20A double-pole switch having a cord outlet and, if desired, a neon indicator.

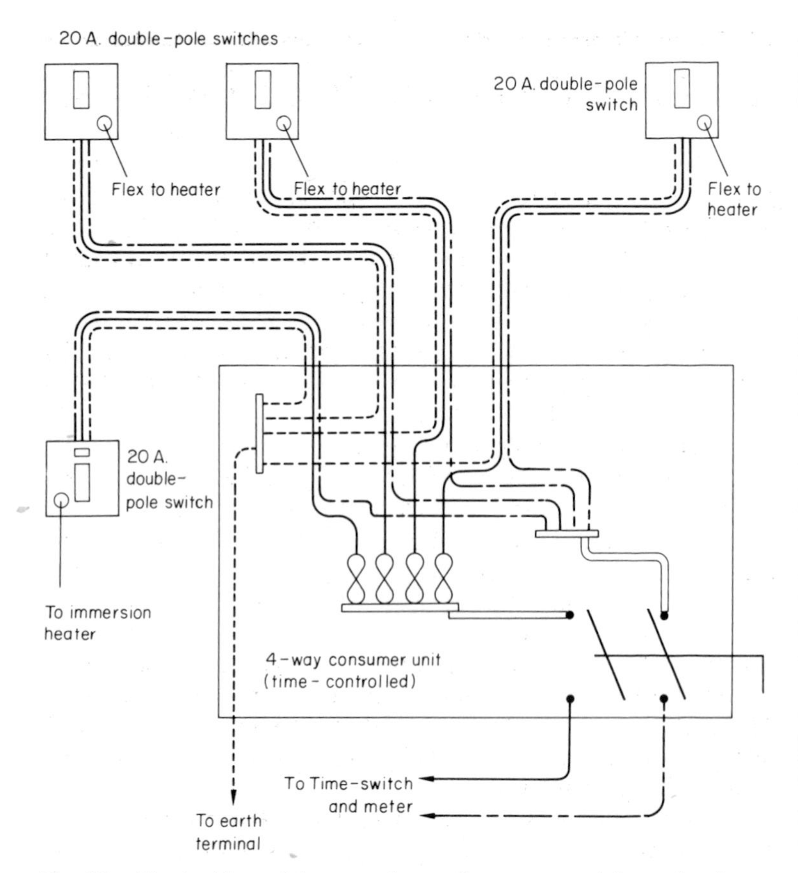

Fig. 87. Circuit wiring and the connections at the consumer unit for an electric storage radiator installation.

Storage heaters are connected to the outlets by sheathed flexible cord. With one type of storage heater – the storage fan heater – a second circuit is required, the two being connected to the storage heater via a linked double switch (see below).

Two principal types of electric storage heater are available for the home: the storage radiator and the storage fan heater.

STORAGE RADIATORS

The storage radiator, the most widely used type, consists of a storage heater block containing electric elements, which are thermally insulated and contained in a steel casing (Fig. 88). Heat is emitted through the steel casing by radiation, its rate of output being controlled by the thickness and grade of thermal insulation. Heat is stored in the heater block during the seven-hour night period when electricity is available in the circuit and is expelled the following day.

Once the heater is charged with heat, the householder has no means of controlling the outflow of heat, but the heater does have an input controller for varying the amount of heat charged and therefore the amount available the next day. Individual heaters can be switched off manually before the charge period commences where heat is not required the following day or for a longer period.

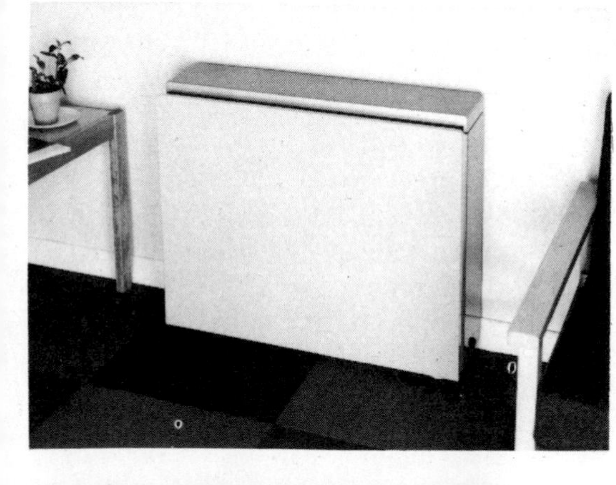

Fig. 88. Storage electric radiator.

Fig. 89. Storage fan heater.

STORAGE FAN HEATERS

The storage fan heater is constructed similarly to the storage radiator but contains a fan which expels the heat from the storage block as warm air (Fig. 89). The thermal insulation is more efficient than in storage radiators and very little heat is lost as radiant heat via the casing. This means that heat is expelled only when the fan is switched on by the user, which can be at any time during the day because the fan receives its energy via a normal 24hr supply.

The circuit for the heater is the same as for a storage radiator, but the fan circuit can be a spur from a ring circuit or even from a lighting circuit since the fan takes very little current. The ring circuit spur is connected to the ring circuit via a fused connection unit and the one spur can supply a number of storage heater fans.

This circuit and the heater circuit from the time-controlled consumer unit are both terminated at a linked switch such as the MK 25A switch designed for the purpose. Two flexible cords are run from this switch to the storage heater terminals, one for the heater section, the other for the fan.

Besides switching the fan on and off manually as heat is required, it is possible to provide automatic switching on and off using a time switch and/or a room thermostat, wired into the fan circuit at the linked switch. A time switch enables the fan to be switched on at a preselected time to warm the room up, whereas a thermostat maintains the required temperature in the room so long as heat is available from the storage heater. For most purposes however manual control of the fan is regarded as sufficient.

12
OUTDOOR WIRING

When mains electricity is required in a detached garage, greenhouse, workshop, shed or other building it is necessary to run a cable from the main switch and meter position in the house and not from a house circuit. The supply in the outbuilding must therefore be independent of the house circuits whether it stands a few feet from the house or at the bottom of a long garden.

Outside lights attached to the house or in the driveway can be, and usually are, supplied from the house lighting circuit. Outdoor socket outlets for the mower, hedge-cutter, or fountain or pool lighting may also be supplied from the house circuits (Fig. 90). A residual current circuit breaker must be provided for all socket outlets used for outdoor equipment.

CABLES TO OUTSIDE BUILDINGS

A cable feeding a detached garage, greenhouse, workshop, shed or summerhouse may be run overhead, underground or along a wall, but *in no circumstances* may it be run along a fence.

For a short distance between the house and outbuilding there is much to be said for running the cable overhead, but for long runs an underground cable is usually preferred; a long overhead cable can be unsightly, and there is always a risk that the cable, or its supports, will be damaged by high winds.

Once laid, an underground cable is out of sight and is unlikely to give any trouble for many years. You do however have the hard work of digging a trench, which is often made more difficult where there is an area of concrete or crazy paving at the side or rear of the house. When there is a wall between garage and house or down the garden this provides an excellent route for the cable, even if a section of it does have to be run underground.

OVERHEAD CABLES

A cable run overhead can be ordinary twin-core and earth

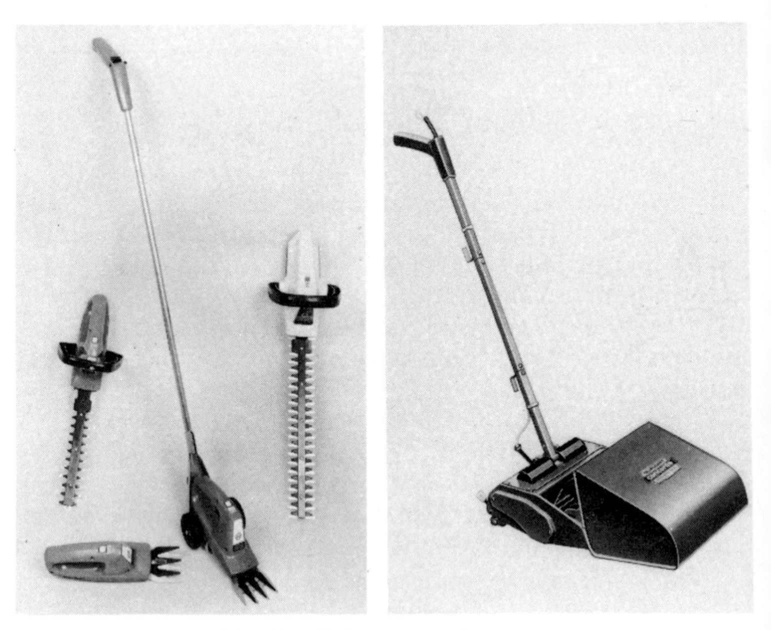

Fig. 90. Hedgecutters and mower.

PVC-sheathed house wiring cable. For a short span (not exceeding 3.7m between the house and outbuilding) no additional support is required for the cable. For longer spans the cable must be attached to or suspended from a catenary wire consisting of galvanized steel wire or cable (see Fig. 91).

The electric cable must be fixed at a minimum height of 3.7m above the general ground level, but above a driveway the minimum height is 5.2m. To attain either of these heights the cable in the house must usually be run up into the attic and under the eaves to the outside. At the outbuilding it is usually necessary to fix a pole or an extension timber upright to attach the cable or its supporting catenary wire.

The catenary wire is attached to eyebolts, but one end requires an adjustable eyelet to take up the slack and so reduce sagging. The cable can be attached to the catenary by insulated clips or be suspended from special slings. Clips or slings should be spaced not more than 225mm apart.

The catenary must be bonded to earth but must not itself be

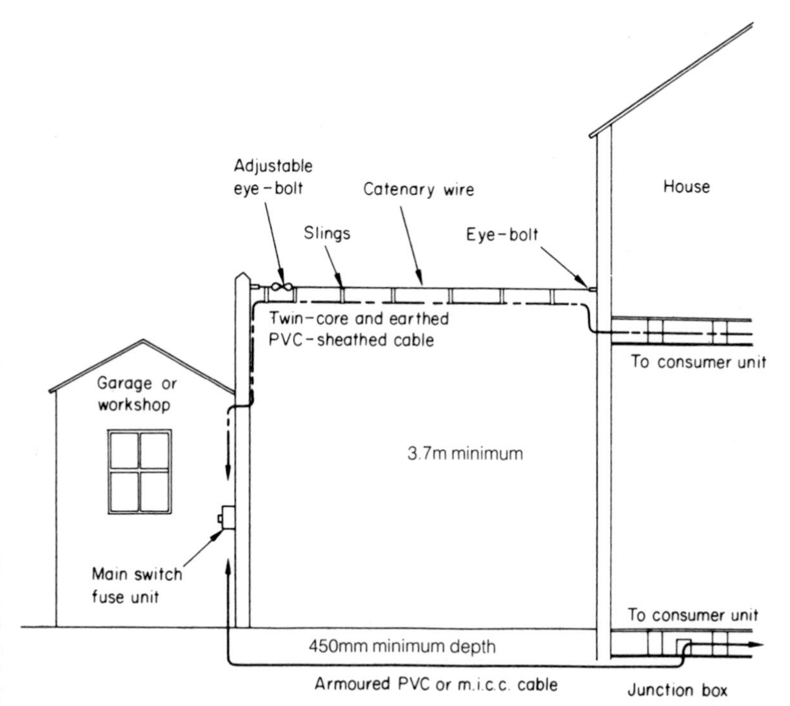

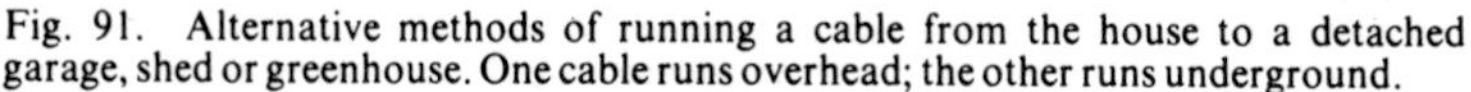
Fig. 91. Alternative methods of running a cable from the house to a detached garage, shed or greenhouse. One cable runs overhead; the other runs underground.

used as the earth continuity conductor (e.c.c.) of the circuit, this being the earth conductor within the sheathed cable.

To bond the catenary wire fix an earthing clip at one end – the outbuilding end is usually the more convenient – and from this clip run a green yellow PVC-insulated earth cable into the building to terminate at the earth terminal of the main switch and fuse unit. The earth conductor within the circuit cable is connected to the same terminal.

UNDERGROUND CABLES

Choose the best route for the cable from the house to the outbuilding and dig a trench at least 450mm deep. If the cable is to pass through a vegetable plot or other cultivated land where deep digging is likely, dig the trench at this section at least 600mm deep (Fig. 91). A preferred route for a cable is at the edge of a path, where the cable is less likely to be disturbed and turf does not have to be lifted. Cables for laying underground, as already stated, may be either armoured PVC/PVC (two-core) with an extruded covering of PVC over the wire armour, or mineral-insulated, copper sheathed (m.i.c.c.) (two-core) with an extruded covering of PVC which is usually orange in colour.

The wire armour of the one cable and the copper sheathing of the other is used as the earth conductor, so there is no earth conductor within the sheath as there is with ordinary PVC-sheathed cable. On the ends of both types of cable metal-screwed glands are fitted for screwing into the cable entry hole of a junction box or the entry hole of metal-clad switch gear. The connection between the gland and the metal box or switch has to make good electrical contact to provide a continuity of the earthing.

M.i.c.c. cable also requires a seal at each end to prevent moisture entering, since the magnesium oxide insulation is hygroscopic quickly absorbs moisture from the atmosphere. For this reason the seals must be fitted as soon as the cable is cut. Fitting a seal requires special tools and some expert knowledge and is best done by the supplier.

Cables cut to the required length can be bought from some shops with seals and glands fitted. The gland is not always necessary, as some switch gear has a cable entry designed to accept an m.i.c.c. seal, but as glands have to be fitted before the seal it is as well to have glands included even if they are not used.

The glands of armoured PVC/PVC cable are simple to fit. Since they are of the compression type they are screwed on to the end of the wire armour using pliers or pipe grips.

The special outdoor cable does not need to run right back to the consumer unit in the house. It is better terminated at a metal junction box fitted between joists underneath the floor-boards near the point where the cable enters the house. From the junction box ordinary two-core and earth PVC-sheathed cable is run to the consumer unit (see Fig. 92).

In the outbuilding the same arrangement can be made, using a metal junction box, provided the main switch or consumer unit is within a few feet of the point where the cable enters and

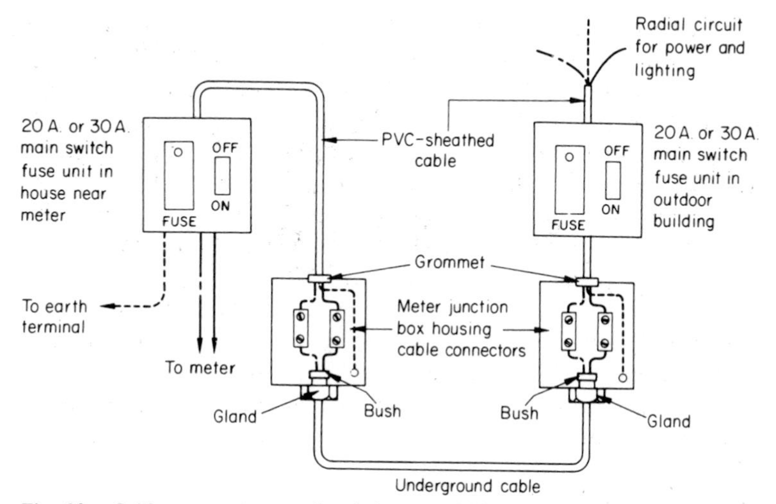

Fig. 92. Cable connections and switch controls for an electricity supply to a detached building. A special distribution unit is often installed in a greenhouse as an alternative to a main switch and fuse unit.

the main switch has a metal casing which will accept a cable gland.

The advantage of armoured cable is that no sealing of ends is required; the advantage of m.i.c.c. cable is that its very small diameter (about that of a pencil) makes it inconspicuous when run on the surface of walls and enables it to be cemented into cracks and brick courses. The price of each is about the same. Provided that it is protected by high impact PVC conduit or galvanised metal conduit, PVC 2-core and earth cable can be run underground.

Lay the cable in the trench, first removing any sharp stones or flints likely to damage the PVC covering and result in corrosion of the armour or copper sheath. If in stony soil place the cable in a few inches of sifted soil or course sand.

Pass one end of the cable through a hole drilled in the house exterior wall above the damp course where it will go into the void beneath the floorboards. If floor is solid the hole for the cable must be cut above skirting level and the junction box fixed to the wall. Under the floor the box is fixed to a piece of 75 × 19mm timber fixed between two joists. Push the other end of the cable through the hole into the outbuilding to a junction box or direct to the switch and fuse unit.

The supply to the outbuilding needs to have a current rating to suit the demand. Normally this will be either a 20A supply using 2.5sq. mm cable or a 30A supply using 4sq. mm cable. For the shed or small workshop requiring a light and a socket outlet, if it is situated fairly near the house a 20A supply will be adequate; but for the garage, greenhouse or well-equipped workshop a 30A circuit is advised, especially where the outbuilding is some distance from the house and requires a long cable and is subject to voltage drop.

For a 20A supply the fuse in the house and the switch and fuse unit will be 20A rating. For a 30A supply these will have a 30A rating. As these socket outlets will most likely be used for outdoor equipment, a residual-current circuit breaker must be installed.

HOUSE END CABLE CONNECTIONS

The outdoor cable within the house is run into the cable outlet of the metal junction box, and the bush or locknut tightened to make good electrical contact between the cable gland and the box.

From this box run twin-core and earth PVC-sheathed cable to the main switch and fuse unit at the consumer unit. At the

Fig. 93. Switch-fuse unit.

junction box the end of this cable is run into a cable outlet in the box fitted with a rubber or PVC grommet. The ends of this cable are prepared and jointed to the ends of the outdoor cable using a terminal block for the two insulated wires. The bare earth wire is enclosed in green yellow PVC sheathing and is connected to an earth terminal in the box. The cover is then fitted to the box.

At the consumer unit the cable can be connected direct to a spare fuse way or to a main switch and fuse unit fixed next to the consumer unit and connected to the meter by the electricity board (Fig. 93). Where the cable is run overhead this cable is run direct to the fuse or main switch unit.

CONNECTIONS WITHIN THE OUTBUILDING

The connections at the main switch and fuse unit in the outbuilding are the same as in the house.

From the switch and fuse unit the interior wiring also is the same as in the house and the same type of socket outlets, switches and other accessories can be used, although if preferred metal clad switches and sockets may be used instead.

For a shed, garage or workshop the circuit may be a 30A radial circuit (p. 79).

In the greenhouse, where there are a number of portable appliances plus fixing wiring, a special distribution control unit is usually fitted. This is equipped with a number of socket outlets and fused connection units, all individually switched and where necessary protected from water spray (see Fig. 94).

OUTDOOR SOCKET OUTLETS

When one or more socket outlets are fixed out of doors to supply an electric mower, hedge-trimmer power drill, or other garden tool, and/or for supplying a fountain pump or pool lighting and other garden lighting, they must be weatherproof and protected by an r.c.c.b.

They can be fixed to the exterior wall of the house or other building, or, if away from buildings and walls, to stout posts of wood or concrete. If desired a small enclosure will protect the socket from the weather.

Wiring for sockets fixed to the house wall can be passed through holes drilled in the wall immediately behind the socket, thereby eliminating an outdoor run of cable. Where the cable is fixed to the exterior wall use m.i.c.c. cable.

For socket outlets remote from buildings and fixed to posts an underground run of cable is needed using the method described above.

13

MISCELLANEOUS PROJECTS

ADDING A LIGHT

When an additional light is required in the house, to be controlled by a switch independently of other lights, a source of electricity must be located. This will usually be at a loop-in ceiling rose or at a joint box.

For this addition use twin-core and earth PVC-sheathed cable of 1.0sq. mm size. Run the cable from the joint box or ceiling rose to a new joint box fitted equidistant from the new

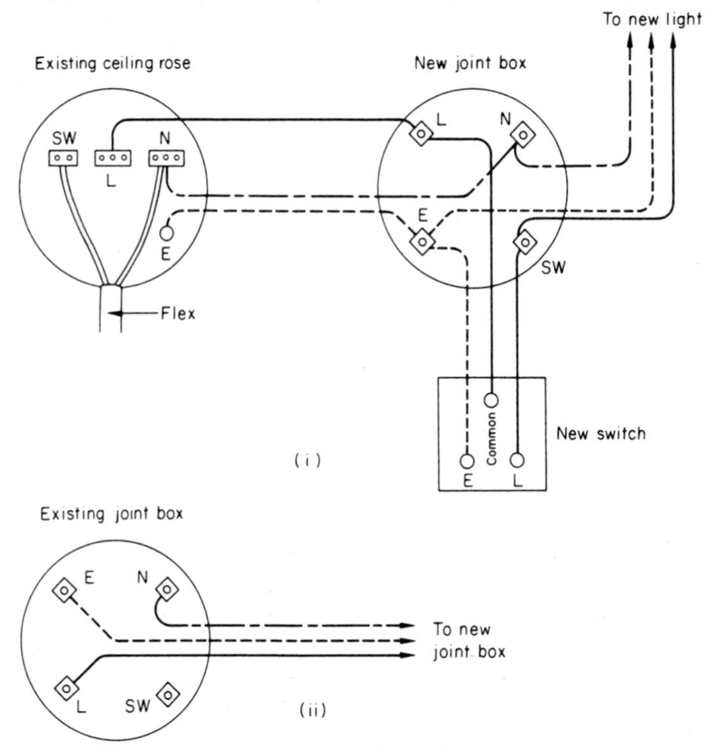

Fig. 94. Circuit and connections when adding a light to an existing lighting circuit: (i) from an existing ceiling rose; (ii) from an existing joint box.

light and switch. From the new joint box run a length of the same cable to the light and another length to the switch (See Fig. 94).

CONVERTING AN ON/OFF SWITCH TO TWO-WAY

Any one-way on/off switch can be converted to two-way switching simply by running a 1.0sq. mm three-core and earth PVC-sheathed cable from the existing switch to the new switch position, replacing the existing one-way switch by a two-way switch and making the connections as in Fig. 95.

ADDING A POWER POINT

An additional 13A power point can usually be supplied from a ring circuit in the form of a spur described on p. 77. Fig. 58.

The cable used is 2.5sq. mm twin-core and earth PVC-

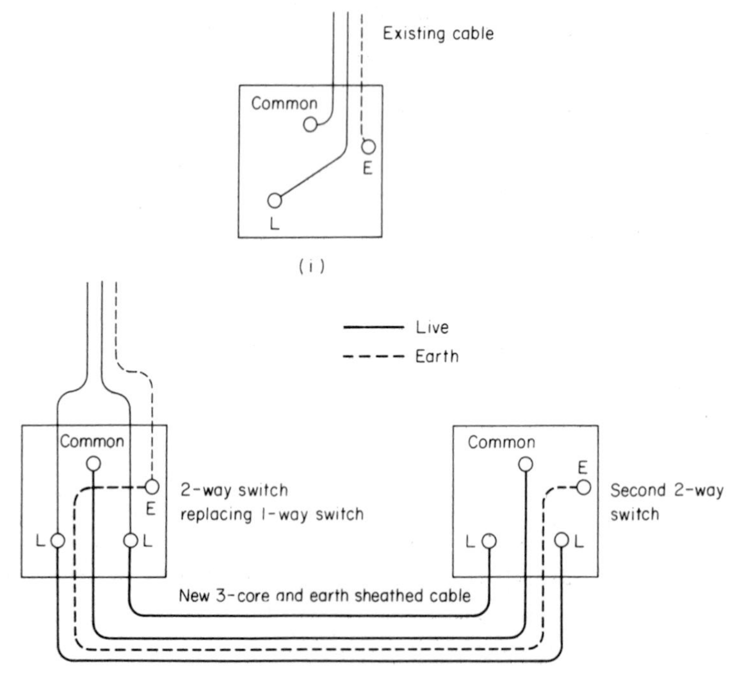

Fig. 95. Converting a one-way switch to two-way switching: (i) existing one-way switch; (ii) wiring for new two-way switch.

Single-switched socket outlet

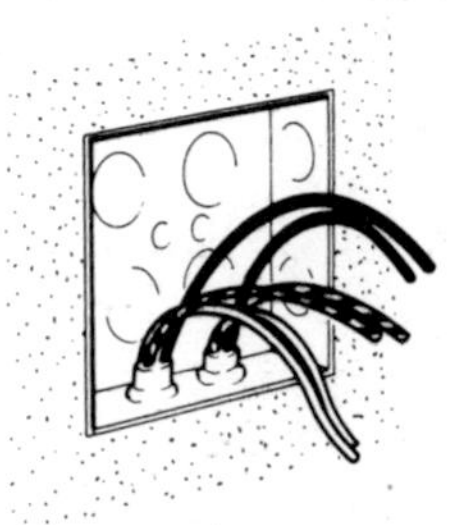

Socket removed showing flush metal box and cables

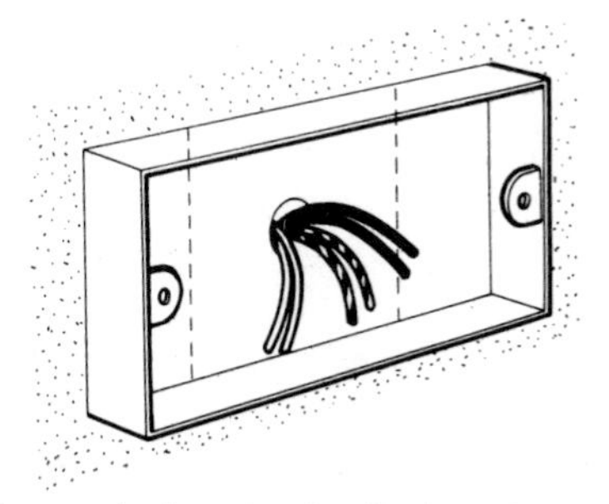

2-gang plastic surface box fixed over 1-gang flush box

Double-socket outlet

Fig. 96. Converting a single flush-mounted 13A socket outlet to a surface-mounted double socket outlet.

sheathed cable. This can be connected to the ring circuit cable at the terminals of a socket-outlet, at the terminals of a fused connection unit or at a 30A three-terminal joint box inserted into the ring cable. A spur cable may supply one single 13A socket outlet or one double socket outlet.

CONVERTING A SINGLE SOCKET OUTLET TO A DOUBLE

Most single 13A socket outlets connected to a ring circuit cable can be converted to double sockets simply by fitting a new two-gang mounting box and replacing the single socket by a double.

When an existing single socket outlet is mounted on a flush metal box sunk into the wall the metal box can be removed and in its place a two-gang flush metal box fitted into a larger chase cut in the wall. Alternatively, a plastic surface two-gang box can be fixed to the single-gang flush box and a double socket fixed to the surface box.

When an existing single socket outlet is surface mounted, simply remove the existing socket outlet and box and replace them by a double socket and two-gang surface box. Fig. 96.

14

MAINTENANCE

MENDING FUSES

When a fuse blows, which may be indicated by failure of lights, lack of power from socket outlets or failure of a cooker, water heater or other appliance to operate, it is first necessary to locate the fuse in the consumer unit or fuse board. If the circuits are not listed in the inside of the cover or if the fuse units are not marked it may be necessary to remove more than one fuse before you find the one that has blown. Colour coding of fuses helps here: for example, if a lighting circuit has failed the fuse will be one of the fuses with white colour coding.

Remove the fuse cover. Examine fuses in turn, according to colour code (in modern units).

A blown fuse of the rewirable type is located by the obviously melted fuse wire, but a 'blown' cartridge fuse cannot be detected by visual inspection because the fuse element is enclosed in the ceramic tube. This must be checked electrically by a suitable device such as a bulb and battery or a metal-cased torch.

With a rewirable fuse, unscrew the terminals of the fuse holder to release the remaining pieces of fuse wire (Fig. 97). Remove any blobs of metal and clean off any marks of burning.

Select from the fuse wire card the wire of the correct current rating. Insert this in the asbestos or ceramic tube and secure the end in the terminal hole or under the terminal screw.

Cut the wire to length and secure the cut end to the other terminal. Finally, trim off the ends.

Before turning off the main switch to insert the fuse holder in the fuse way, try to ascertain why the fuse blew. If, for example, it happened when a particular light was switched on this would indicate a fault in the fitting, probably a worn flex or even a dud lamp.

Insert the fuse and turn on the main switch. If the fuse immediately blows again a fault is indicated and it may be necessary to call in an electrician. If it blows after an hour or so it is

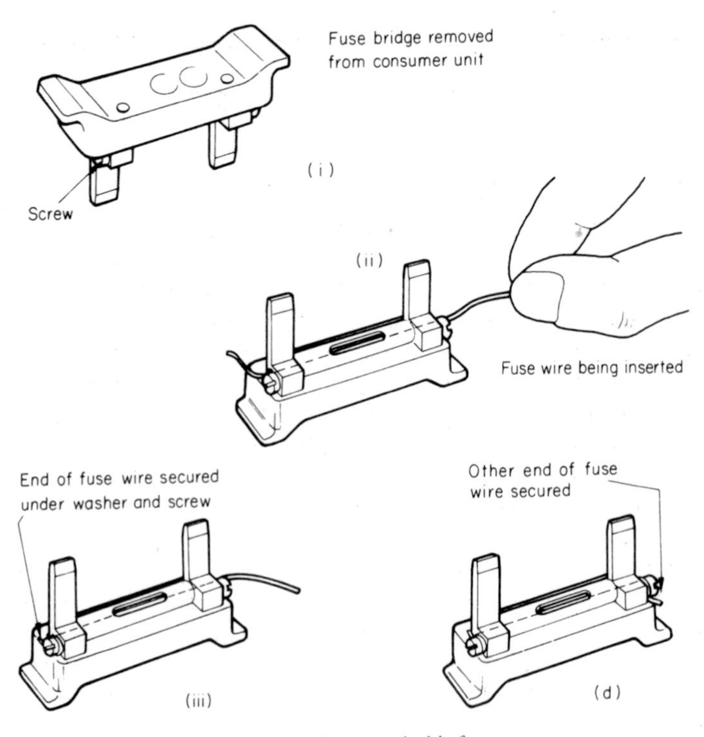

Fig. 97. Mending a rewirable fuse.

probably an overload.

To mend a cartridge fuse, remove the fuse holder and push out the fuse cartridge. Test to check that it is a dud. Discard the cartridge and select a new cartridge fuse of the same colour and current rating (Fig. 98). Insert this in the fuse holder and turn on the main switch. If it blows again the symptoms are the same as a rewirable fuse and the action to take is the same.

Where a consumer unit or fuse board is fitted with cartridge fuses a couple of spares of each current rating should be kept at hand, as should a card of fuse wire for rewirable fuses. Some consumer units have a container for cartridge fuses, but some do not.

TESTING OF WIRING BY ELECTRICITY BOARD

An electricity board will usually test an installation before con-

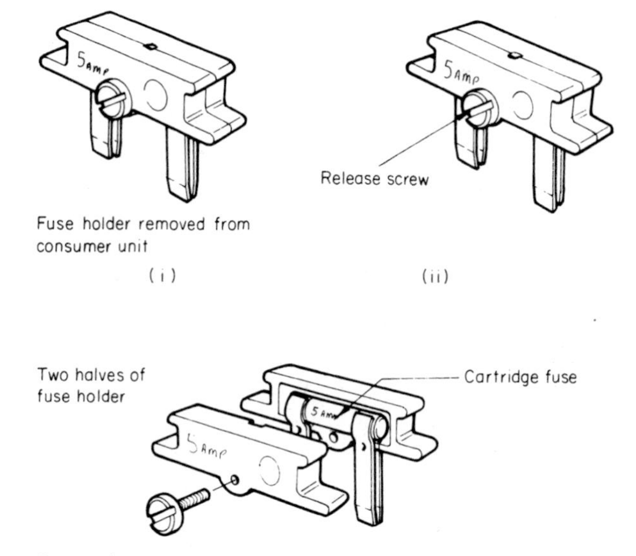

Fig. 98. Mending a cartridge fuse.

necting it to the mains supply. It does this for a new installation and where major alterations have been made to an existing installation such as rewiring or the installation of a ring circuit.

The board may also test an installation upon a change of consumer, that is when you take over a dwelling and apply for a supply of electricity. This is usually termed a re-connection.

If the wiring when tested is seen to be faulty or badly carried out the board may refuse to connect it until faults or bad work have been rectified. No charge is made for the first test but a charge is made for necessary subsequent tests.

In the case of a reconnection the board may reconnect the installation if it is not up to standard but is not dangerous; this is however only a temporary connection until the faulty work is rectified, following a test report from the district engineer.

So long as the work is carried out in compliance with the IEE Wiring Regulations the board will connect both new and existing wiring.

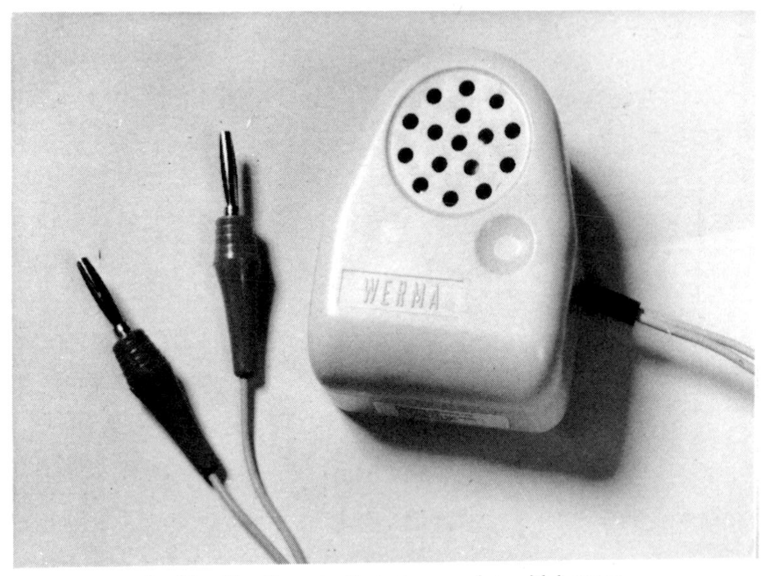

Fig. 99. Small continuity tester complete with battery.

SAFETY IN WIRING

In the interests of safety it is essential that all wiring conforms with the requirements of the IEE Wiring Regulations with regard to both the quality of the materials used and the workmanship. The advice given in this book is in accordance with these regulations, but the choice of materials and the workmanship lies with the individual.

Of special importance are the following rules:

– Do not attempt jobs beyond your capability or knowledge.

– Always turn off the current at the main switch before commencing any job and where necessary remove fuses in case someone else turns on the main switch unaware that you are working on the installation.

– When carrying out tests or locating faults requiring the current to be switched on and off, take special care that the current is off when you handle live wires or contacts.

– Always use the correct sizes and types of cables.

– Do not install a socket outlet, switch or other wiring accessory without the necessary mounting box or pattress.

– Never install socket outlets in a bathroom or make any provision for using portable mains-operated appliances in the bathroom except a shaver from the approved type of shaver supply unit. For instance, do not plug appliances into lampholders of lighting fittings.

– Do not supply an immersion heater from a plug and socket outlet where the tank cupboard opens into the bathroom and therefore could be used for portable appliances.

– Do not take portable appliances into the bathroom supplied from a socket outlet situated outside the bathroom, e.g. on the landing.

– Make certain that no switch except the insulated cord of a cord-operated ceiling switch can be reached by anyone using the bath or shower. This includes a switch for the immersion heater in the tank cupboard.

– Do not run a number of portable appliances (or fixed appliances) from one socket outlet.

– Keep flexible cords as short as possible and do not run flexes under carpets, rugs or other floor coverings.

– Do not carry out extensions with flexible cord.

– Do not make temporary 'hook-ups', for these tend to become permanent and are a potential danger even for short periods.

– Do not increase the current ratings of fuses by using larger-size fuse wire.

– Do not attempt to rewire cartridge fuses or use substitute materials where no spare cartridge fuses are immediately available.

– Always ensure that the earthing arrangements are effective.

– Do not use appliances and electrical tools out of doors or in the greenhouse unless they are designed for these situations.

– Do not use appliances that are not earthed unless they are double-insulated, (marked with a double hollow square) and therefore do not require earthing.

IMPERIAL TO METRIC

With the growing use in the electrical industry the following table will help readers to convert Imperial linear measurements to metric when necessary.

IMPERIAL in	METRIC mm	IMPERIAL ft	METRIC metres (m)
$\frac{1}{16}$	1.6	3	0.9144
$\frac{1}{8}$	3.2	1 yd	0.914
$\frac{3}{16}$	4.8	4	1.219
$\frac{1}{4}$	6.4	5	1.524
$\frac{5}{16}$	7.9	(2 yd) 6	1.829
$\frac{3}{8}$	9.5	7	2.135
$\frac{1}{2}$	12.7	8	2.440
$\frac{9}{16}$	14.3	(3 yd) 9	2.745
$\frac{5}{8}$	15.9	10	3.050
$\frac{11}{16}$	17.5	11	3.350
$\frac{3}{4}$	19.1	(4 yd) 12	3.660
$\frac{13}{16}$	20.6	13	3.960
$\frac{15}{16}$	23.8	14	4.270
1	25.4	(5 yd) 15	4.750
$1\frac{1}{4}$	31.8	16	4.880
$1\frac{1}{2}$	38.1	17	5.180
$1\frac{3}{4}$	44.5	(6 yd) 18	5.485
2	57.2	19	5.790
$2\frac{1}{2}$	63.5	20	6.095
3	72.2	(10 yd) 30	9.145
4	101.6	40	12.190
5	127.0	50	15.240
6	152.4	(25 yd) 75	22.860
7	177.8	100	30.480
8	203.2	200	60.960
9	228.6	(100 yd) 300	91.440
10	254.0	500	152.400
11	279.4	1000	304.800
12	304.8		
15	381.0		
16	406.4		
18	457.2		
21	535.4		
24	609.6		
27	685.8		